Analytic Functions

PRINCETON MATHEMATICAL SERIES

Editors: MARSTON MORSE and A. W. TUCKER

1. The Classical Groups, Their Invariants and Representatives. By HERMANN WEYL.
2. Topological Groups. By L. PONTRJAGIN. Translated by EMMA LEHMER.
3. An Introduction to Differential Geometry with Use of the Tensor Calculus. By LUTHER PFAHLER EISENHART.
4. Dimension Theory. By WITOLD HUREWICZ and HENRY WALLMAN.
5. The Analytic Foundations of Celestial Mechanics. By AUREL WINTNER.
6. The Laplace Transform. By DAVID VERNON WIDDER.
7. Integration. By EDWARD JAMES McSHANE.
8. Theory of Lie Groups: I. By CLAUDE CHEVALLEY.
9. Mathematical Methods of Statistics. By HARALD CRAMÉR.
10. Several Complex Variables. By SALOMON BOCHNER and WILLIAM TED MARTIN.
11. Introduction to Topology. By SOLOMON LEFSCHETZ.
12. Algebraic Geometry and Topology. Edited by R. H. FOX, D. C. SPENCER, and A. W. TUCKER.
13. Algebraic Curves. By ROBERT J. WALKER.
14. The Topology of Fibre Bundles. By NORMAN STEENROD.
15. Foundations of Algebraic Topology. By SAMUEL EILENBERG and NORMAN STEENROD.
16. Functionals of Finite Riemann Surfaces. By MENAHEM SCHIFFER and DONALD C. SPENCER.
17. Introduction to Mathematical Logic, Vol. I. By ALONZO CHURCH.
18. Algebraic Geometry. By SOLOMON LEFSCHETZ.
19. Homological Algebra. By HENRI CARTAN and SAMUEL EILENBERG.
20. The Convolution Transform. By I. I. HIRSÇHMAN and D. V. WIDDER.
21. Geometric Integration Theory. By HASSLER WHITNEY.
22. Qualitative Theory of Differential Equations. By V. V. NEMICKII and V. V. STEPANOV. Translated by SOLOMON LEFSCHETZ. (In press).
23. Topological Analysis. By GORDON T. WHYBURN.
24. Analytic Functions. By L. AHLFORS, H. BEHNKE and H. GRAUERT, L. BERS, et al.
25. Continuous Geometry. By JOHN VON NEUMANN.
26. Riemann Surfaces. By LARS V. AHLFORS and LEO SARIO.

Analytic Functions

By

R. NEVANLINNA

H. BEHNKE and H. GRAUERT

L. V. AHLFORS · D. C. SPENCER

L. BERS · K. KODAIRA

M. HEINS · J. A. JENKINS

PRINCETON, NEW JERSEY

PRINCETON UNIVERSITY PRESS

1960

L.C. CARD: 59-11081

PRINTED IN THE UNITED STATES OF AMERICA

Preface

THIS book contains the principal addresses delivered at the Conference on Analytic Functions held September 2–14, 1957 at the Institute for Advanced Study, Princeton, New Jersey. Besides these addresses, some fifty shorter papers were presented in the five seminars into which the Conference was divided. These shorter papers, which are listed by author and title in the Appendix, will appear under the title "Seminars on Analytic Functions" (Vol. 1, 2) in photo-offset edition, published by the Air Force Office of Scientific Research (ARDC) and distributed by the Institute for Advanced Study to members of the Conference, selected libraries, and a distribution list supplied by the Air Force.

The Conference was called by the Institute for Advanced Study under Contract No. AF 18(603)–118 with the Air Force Office of Scientific Research. Acknowledgement is hereby made of the cooperation of Dr. Carl Kaplan, Chief of Scientific Research, Air Force Office of Scientific Research; Dr. Merle Andrew, Chief, Mathematics Division, Air Force Office of Scientific Research; Dr. Robert Oppenheimer, Director, The Institute for Advanced Study; Mr. Minot C. Morgan, Jr., General Manager, The Institute for Advanced Study; and of the Princeton University Press as publisher of the present volume.

Princeton, New Jersey
March 17, 1958

MARSTON MORSE, *Chairman*
ARNE BEURLING
ATLE SELBERG
Organizing Committee of the Conference

Contents

Analytic Functions

On Differentiable Mappings

Rolf Nevanlinna
ACADEMY OF FINLAND

1. Introduction

THE treatment of the problem of mappings in linear spaces can be based formally as well as intuitively on an infinitesimal calculus which is co-ordinate-free and dimension-free. Such a differential calculus has been introduced by Fréchet for infinite-dimensional spaces, and has been applied by him and others after him (Graves, Hildebrandt, Taylor, Rothe, Lusternik, Sobolev, etc.), principally in functional analysis. In the course of the last five years my brother and I have developed systematically such an infinitesimal technique in lectures in differential geometry, in the theory of manifolds and partial differential equations, and in the calculus of variations [1].

Thus, I have become convinced that such a method, which is directly connected with the fundamental conceptions of Leibniz in the differential and integral calculus, brings a greater simplicity and unity not only into functional analysis but also to the elementary classical questions in analysis of a *finite* number of variables. As one example I need mention only the tensor calculus: on the one hand, the typographically awkward indices are avoided; on the other, one does not go so far in the direction of abstract concepts (as in the case of Bourbaki) that one loses the means of what I call an "automatic calculating" technique.

The advantages of such an "absolute" calculus lie, above all, in three directions: (1) The notation and symbols, which have become standard in the infinitesimal calculus of a real function of one real variable ever since the time of Leibniz, remain unchanged in the most general cases; the calculus is "dimension-free." (2) Because of the "coordinate-freedom," many of the invariance proofs are superfluous. (3) The calculus emphasizes the invariant geometric intuition and thereby assists in the discovery of natural problems and methods of proof.

In what follows I want to apply what I have just said to a classical problem which lies within the scope of this conference: I shall give an elementary direct proof of Liouville's theorem which states that every conformal transformation in Euclidean space of dimension greater than two must necessarily reduce to a translation, a magnification, an orthogonal transformation, a reflection through reciprocal radii, or a combination of these elementary transformations [2].

2. Differentials

For the proof we shall recall some simple notations and concepts, as follows.

We consider two linear spaces $\mathfrak{R}_x$ and $\mathfrak{R}_y$ of finite or infinite dimension and with real structure. Suppose that there exists a complete metric in each space, either in the sense of Euclid-Hilbert or of Minkowski-Banach. In the former case we denote, as usual, by (h, k) the inner product of two vectors h, k belonging to $\mathfrak{R}_x$ or else to $\mathfrak{R}_y$. For the norm of a vector x or y we shall use the notations $|x|$ or $|y|$, respectively.

Let $y = y(x)$ be a function which maps a domain $\mathfrak{G}_x$, contained in $\mathfrak{R}_x$, into the space $\mathfrak{R}_y$. This function is defined to be differentiable at the point $x \in \mathfrak{G}_x$, if there exists a bounded linear transformation $A(x)$ of $\mathfrak{G}_x$ into $\mathfrak{R}_y$ with the following property. The difference $\Delta y = y(x + \Delta x) - y(x)$ is given in the first approximation by the vector $A(x)\Delta x$, the image under the transformation $A(x)$ of the vector Δx:

$$\Delta y = A\Delta x + |\Delta x| \cdot \varepsilon(\Delta x),$$

where ε is a vector in $\mathfrak{R}_y$ which vanishes for $|\Delta x| \to 0$. The first term on the right is defined as the differential $dy(x)$ of $y(x)$, and the operator $A = A(x)$ as the derivative of $y(x)$. Because $\Delta x = dx$ for $y = x$, we may also write $dy = Adx$. For the derivative we use the Lagrange notation $A = y'$ or the symbol $A = \dfrac{dy}{dx}$ of Leibniz.

Consider the differential $dy = y'(x)h$, $(h = dx)$. If we keep h constant and put $dx = k$, then we obtain the second differential as:

$$d^2y = d(y'h) = y''kh.$$

This is a bilinear function of the independent differentials $dx = h$ and $dx = k$. By definition the operator $y'' = \dfrac{d^2y}{dx^2}$ is the second derivative of y at x.

The differential d^ny of order n is similarly defined as a multi-linear form depending linearly on n independent differentials $d_1x, \ldots, d_nx$:

$$d^ny = y^{(n)}(x)d_1x \ldots d_nx.$$

The operator $y^{(n)} = \dfrac{d^ny}{dx^n}$ is the derivative of y of order n.

3. Basic theorems

I shall briefly indicate how some elementary theorems of differential calculus can be expressed in terms of these notations.

1. Commutativity of differentiation. The second derivative is symmetric: $y''hk = y''kh$.

This can be proved under the condition that the operator y'' *exists*; continuity is not required. This property, however, is not equivalent to the classical statements found in all textbooks.

2. Mean-value theorem. This is nothing but the classical mean-value theorem, formulated as follows. We consider again the mapping $y = y(x)$ and denote as above $\Delta y = y(x + \Delta x) - y(x)$. Let $\mathfrak{R}_{y*}$ be the dual space of $\mathfrak{R}_y$, defined by all real linear functionals y^* on $\mathfrak{R}_y$. Take an arbitrary functional y^* and form the real function $f(\tau) = y^*y(x + \tau\Delta x)$ $(0 \leq \tau \leq 1)$. If y' exists for $0 < \tau < 1$ the elementary mean-value theorem gives:

$$y^*\Delta y = y^*dy(\xi),$$

where $\xi = x + \tau \cdot \Delta x$ $(0 < \tau < 1)$; this value ξ depends on y^*. If $\mathfrak{R}_y$ is a Euclidean or a Hilbert space we choose $y^*y = (y, e)$, where e is an arbitrary constant vector. In this special case the mean-value theorem becomes $(\Delta y, e) = (dy(\xi_e), e)$.

3. Implicit functions. The basic theorem about implicit functions can be proved by the above mean-value theorem. We postulate:

(a) The function $y = y(x)$ is differentiable for $|x| < r_x$.

(b) The lower bound:

$$\inf_{|h|=1} |y'(0)h| = m$$

is positive, and $y'(0)$ defines an isomorphism between $\mathfrak{R}_x$ and $\mathfrak{R}_y$.

(c) The norm of the transformation $y'(x) - y'(0)$ satisfies the relation:

$$\sup_{|x|\leq r} |y'(x) - y'(0)| \leq \varepsilon(r),$$

where $\varepsilon(r) \leq m$ for $r < r_x$.

Under these conditions the mapping $x \to y$ is one-to-one for $|x| < r_x$, and the image contains the whole sphere $|y| < r_y$, where

$$r_y = \int_0^{r_x - 0} (m = \varepsilon(r))dr \geqq \int_{r=0}^{r_x - 0} r d\varepsilon(r).$$

This statement was recently given by F. Nevanlinna [3]. The value of r_y is the best possible in the following sense: Let $\varepsilon(r) \leq m$ be given as an arbitrary increasing function in the interval $0 \leq r \leq r_x$. Then there exists an extremal function $y(x)$ satisfying the conditions (a), (b), (c) and which maps the sphere $|x| < r_x$ one-to-one, in such a manner that the largest sphere contained in the image has exactly the radius r_y.

4. Liouville's theorem

Let $\mathfrak{R}_x$ be an n-dimensional Euclidean or a Hilbert space ($n \leq \infty$). We consider again a vector function $y = y(x)$ which maps the sphere $\mathfrak{G}_x : |x| < r_x$ into the space $\mathfrak{R}_y (\equiv \mathfrak{R}_x)$ $(y(0) = 0)$. We suppose $y(x)$ continuously differentiable in $\mathfrak{G}_x$. If we assume moreover that $y'(0)$ is an automorphism of $\mathfrak{R}_x$, we conclude (according to the theorem of §3) that the radius r_x can be taken so small that the mapping $\mathfrak{G}_x \rightarrow \mathfrak{R}_y$ is one-to-one.

In order to prove the Liouville theorem we now suppose:
A. The dimension $n > 2$; B. The mapping $x \leftrightarrow y$ is conformal; C. The derivatives $y', \ldots, y^{(4)}$ exist.

By B there is a positive function $\lambda(x)$ such that $|dy(x)| = \lambda(x) \cdot |dx|$, and we get by polarization:

(1) $$(y'h, y'k) = \lambda^2 \cdot (h, k),$$

for every pair of vectors h, k.

We fix three orthogonal vectors h, k, ℓ. Then the vectors $y'h$, $y'k$, $y'\ell$ also are orthogonal. By differentiation of the relation $(y'h, y'k) = 0$ with respect to the differential $dx = \ell$ we obtain:

$$(y''\ell h, y'k) + (y'h, y''\ell k) = 0,$$

and it follows by a cyclic permutation that $(y''hk, y'\ell) = 0$. The differential $y''hk$ is, consequently, orthogonal to $y'\ell$, thus to the orthogonal complement of the plane spanned by the two vectors $y'h$, $y'k$. Therefore $y''hk$ lies in this plane, and we get a decomposition:

$$y''hk = \mu \cdot y'h + \nu \cdot y'k.$$

Here,

$$\mu = \frac{(y''kh, y'h)}{|y'h|^2} = \frac{\lambda'k}{\lambda}, \quad \nu = \frac{\lambda'h}{\lambda}.$$

If we put $\rho = \dfrac{1}{\lambda}$ this formula becomes:

$$\rho'k \cdot y'h + \rho'h \cdot y'k + \rho \cdot y''hk = 0.$$

This holds for every pair of orthogonal vectors h, k.

By differentiation with respect to $dx = \ell$ we obtain:

$$\rho''\ell k \cdot y'h + \rho'k \cdot y''\ell h + \rho''\ell h \cdot y'k + \rho'h \cdot y''\ell k$$
$$+ \rho'\ell \cdot y''hk + \rho \cdot y'''\ell hk = 0.$$

Here the sum of the five last terms is symmetric in ℓ and h. Thus the same property holds for the first term, and we get $\rho''\ell k \cdot y'h = \rho''hk \cdot y'\ell$. But $y'h$ and $y'\ell$ are independent, and hence $\rho''hk = 0$ whenever $(h, k) = 0$.

Now it is an elementary fact, very easy to prove, that a real symmetric bilinear form $\rho''hk$, which vanishes for all orthogonal vectors h, k, has necessarily the form:†

$$\rho''hk = \sigma \cdot (h, k).$$

In our case the real multiplier σ is, *a priori*, a function of the point x, but we shall show that σ must be constant.

In fact, by differentiation we obtain:

$$\rho'''\ell hk = \sigma'\ell \cdot (h, k),$$

and because ρ''' is symmetric,

$$0 = \sigma'\ell \cdot (h, k) - \sigma'h \cdot (\ell, k) = (\sigma'\ell \cdot h - \sigma'h \cdot \ell, k).$$

This holds for every k, and it follows that $\sigma'\ell \cdot h - \sigma'h \cdot \ell = 0$ and (since h and ℓ are independent) $\sigma'h = \sigma'\ell = 0$, $\sigma = \text{const}$.

By integrating $\rho''hk = \sigma \cdot (h, k)\,(dx = h)$ we obtain $\rho'k = \sigma\,(x - x_0, k)$, and a second integration $(k = dx)$ gives:

$$\rho(x) = \frac{1}{\lambda(x)} = \alpha|x - x_0|^2 + \beta,$$

where x_0 is a constant vector and α $(= \sigma/2)$, β real constants. Here $\alpha^2 + \beta^2 > 0$; but we can show that either $\alpha = 0$ or $\beta = 0$.

For the proof, consider the inverse mapping $x = x(y)$. We obtain, as above:

$$\lambda(x) = \frac{1}{\rho(x)} = \gamma|y - y_0|^2 + \delta.$$

Consequently,

$$(\alpha|x - x_0|^2 + \beta)(\gamma|y - y_0|^2 + \delta) = 1. \tag{2}$$

Thus, the images of the points of the sphere $|x - x_0| = r$ lie on the sphere $|y - y_0| = r'$, where the radii r and r' are bounded by the relation $(\alpha r^2 + \beta)(\gamma r'^2 + \delta) = 1$.

Let now $x = a$ $(|a| = 1)$, be a constant vector such that $x_0 = \tau_0 \cdot a$. Putting $x = \tau \cdot a$, $dx = d\tau \cdot a$, dx is orthogonal to the sphere $|x - x_0| = r$

† Let Bhk be a real symmetric bilinear form which vanishes for $(h, k) = 0$. For a fixed $h \neq 0$ the linear form $Lk = Bhk - \frac{Bhh}{(h, h)}(h, k)$ vanishes for $k = h$ and for every k orthogonal to h. Thus $Lk = 0$. If $k \neq 0$ we obtain similarly $Bkh - \frac{Bkk}{(k, k)}(k, h) = 0$. It follows that $\frac{Bhh}{(h, h)} = \frac{Bkk}{(k, k)}$ is a constant σ, and $Bhk = \sigma \cdot (h, k)$.

and the corresponding differential dy is orthogonal to $|y - y_0| = r'$. By integration it follows that the segment $0x$ corresponds to a segment $0y$ $(y = y(x))$ which is radial to the sphere $|y - y_0| = r'$. We obtain therefore:

$$|y| = \int_0^y |dy| = \int_0^x \lambda \cdot |dx| = \int_0^\tau \frac{d\tau}{\alpha\tau^2 + \beta}.$$

If $\alpha \neq 0$, $\beta \neq 0$, the last integral is a transcendental function of $|x|$ (a logarithm or an arc tan). But the relation (2) shows that this function must be algebraic. The solution of this contradiction is that either $\alpha = 0$ or $\beta = 0$.

Suppose first $\alpha = 0$. In this case we have $\lambda = \frac{1}{\rho} = \text{const.}$ By differentiation of the relation $(y'h, y'h) = \lambda^2 \cdot (h, h)$ we obtain for an arbitrary k, $(y''kh, y'h) = 0$. It follows that $y''kh = 0$. The integral of this differential is $y'h = Uh$, where U is a linear transformation of $\mathfrak{R}_x$ into itself. But the equation (1) shows that U has the form λT, where the transformation T is orthogonal. Putting $h = dx$, we obtain finally:

$$y = \lambda Tx + \text{const},$$

and the mapping $x \to y$ is a Euclidean motion, combined with a magnification.

Finally, if $\beta = 0$, $|dy| = \frac{|dx|}{\alpha|x - x_0|^2}$, we transform the space $\mathfrak{R}_x$ by reciprocal radii:

$$x^* = \frac{x - x_0}{|x - x_0|^2},$$

and we get $|dy| = \lambda^* \cdot |dx^*|$, where λ^* is constant. By this transformation, the case $\beta = 0$ reduces to the first case $\alpha = 0$, and the proof of Liouville's theorem is achieved.

5.

An open question is whether this result holds assuming the existence of the first derivative y' only.

Finally, we remark that this proof works also in the case of an indefinite fundamental form (h, k), which is not degenerate. If the dimension of $\mathfrak{R}_x$ is infinite this further requires that this form is what I have called bounded [4]. This means that there exists a dominating complete Hilbert metric $(h, k)_H$ such that $|(h, h)| \leq (h, h)_H$. Under this condition, the above proof remains valid with some modifications.

References

[1] See for instance: Bemerkung zur Funktionalanalysis (*Math. Scand.* **1**, 1 (1953)); Bemerkung zur absoluten Analysis (*Ann. Ac. Sci. Fenn. A. I,*

169 (1954)); Über die Umkehrung differenzierbarer Abbildungen (*ibid. A. I*, 185 (1955)).

[2] The following proof is a simplification of a proof given in the paper: Die konformen Selbstabbildungen des euklidischen Raumes (*Revue la de Fac. des Sci. d'Istanbul* **19**, 3 (1954)).

[3] F. Nevanlinna, Über die Umkehrung differenzierbarer Abbildungen (*Ann. Ac. Sci. Fenn. A. I*, 245 (1957)).

[4] See: Über metrische lineare Räume III (*Ann. Ac. Sci. Fenn. I, A.* 115 (1952)) and: Erweiterung der Theorie des Hilbertschen Raumes (*Comm. Sém. Math. Lund dédié à M. Riesz* (1952)).

Analysis in Non-Compact Complex Spaces†

H. Behnke and H. Grauert

MÜNSTER, WESTFALEN

ABOUT one hundred years have elapsed since the recognition that as domains of existence of analytic functions of a complex variable, the Riemann surfaces appear. The analogue of Riemann surfaces for analytic functions of several complex variables—that is, the complex spaces—has, however, become known only in the past few years. This, of course, is not accidental. Already the first specific results in the field of function theory of several complex variables, discovered by Henri Poincaré, Pierre Cousin, and Fritz Hartogs at the turn of the century, showed that a function theory of several complex variables must necessarily be far more complicated than the classical theory. For the mastery of analytic functions of several complex variables, completely new concepts, just created under the influence of present-day topology and the modern French school, became necessary. These concepts are, for example: complex manifolds, complex spaces, analytic sheaves and fibre bundles, modifications, analytic projections, analytic decompositions, etc.

An introduction to the theory of complex spaces and their functions will now be given. A few historical details must be stated for reference. In 1905 Hartogs [1] discovered that there are certain domains in the space of two complex variables z_1, z_2, such that those functions which are holomorphic in these domains are holomorphic simultaneously in larger domains.

Let $f(z_1, z_2)$ be holomorphic for $z_1 = 0$, $|z_2| < 1$ and for $|z_1| < 1$, $|z_2| = 1$. Then $f(z_1, z_2)$ is holomorphic in the whole dicylinder: $|z_1| < 1$, $|z_2| < 1$.

† Delivered by H. Behnke.

This can be clearly seen in the accompanying figure.

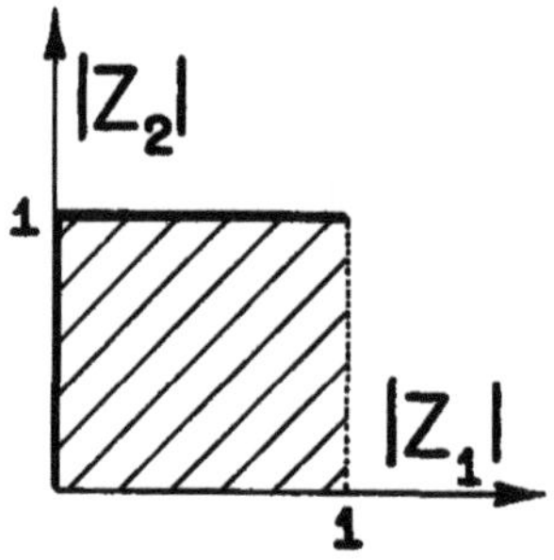

This leads immediately to the much discussed theorem: If $f(z_1, z_2)$ is holomorphic on the boundary of the hypersphere $|z_1|^2 + |z_2|^2 \leq 1$, then $f(z_1, z_2)$ is holomorphic everywhere in the interior $|z_1|^2 + |z_2|^2 < 1$.

Three decades later Peter Thullen added to this a general theory [2]. His main results can be summarized in a few words. Let C^n be the space of n complex variables, G a domain of C^n, $\mathfrak{H}(G)$ the envelope of holomorphy of G—that is, the largest domain containing G in which all holomorphic functions in G are still holomorphic. G is called a *domain of holomorphy* if there is a function holomorphic in G that cannot be holomorphically continued beyond G. The theorem of Thullen states that $\mathfrak{H}(G)$ is a domain of holomorphy and, consequently, the smallest domain of holomorphy containing G. Since, in general, not every domain of C^n is a domain of holomorphy, it follows that, in general, $\mathfrak{H}(G)$ is different from G. Immediately following this, Peter Thullen and Henri Cartan gave examples of domains whose envelopes of holomorphy are not schlicht. Such a domain in C^2 is, for example:

$$G : \tfrac{1}{2} < |z_1| < 1, \quad -\frac{\pi}{2} + \arg z_1 < |z_2| < +\frac{\pi}{2} + \arg z_1; \qquad 0 \leq \arg z_1 \leq 3\pi.$$

Then $\mathfrak{H}(G)$ is:

$$\tfrac{1}{2} < |z_1| < 1; \quad |z_2| < \frac{\pi}{2} + \arg z_1 \quad \text{for} \quad 0 \leq \arg z_1 < 2\pi,$$

and lying over this:

$$|z_2| < \frac{\pi}{2} + \arg z_1 \quad \text{for} \quad 2\pi \leq \arg z_1 \leq 3\pi.$$

This surprising result, which directly followed the discovery in the Thullen theorem, caused next the *ad hoc* construction of the non-schlicht domains over the complex number space C^n. This was done by an appropriate linking of polycylinders. Yet, a year later in H. Behnke and P. Thullen's joint report [3] on function theory of several complex variables, the construction of the non-schlicht domains is presented following the classical example of Hermann Weyl's *Idee der Riemannschen Fläche*. However, according to the stage of function theory and topology at that time, essential additional assumptions have to be made. Thus all ramification points are left unconsidered, whether they have uniformizable neighborhoods or not.

In modern terminology a domain, in the sense of this report, is a complex manifold of complex dimension n lying without ramification points over the projectively closed number space $\bar{C}^n$. (Local coordinates are therefore $z_1, \ldots, z_n$ or, respectively, their projective images.) For almost twenty years this concept of the concrete (Riemann) domain has sufficed for the research papers in the field of function theory. Even Oka in 1954 used this concept of a domain in his paper [4], in which he gave the proof, in a very general case, of E. E. Levi's important conjecture about the characterizations of domains of holomorphy by means of local properties of the boundary.

1. Complex spaces

In the meantime the topologists had successfully introduced the abstract complex manifold. About 1950 began the function-theoretic investigations of complex manifolds. H. Cartan, Serre, Stein, and Behnke presented results of these investigations at the symposium in Brussels in March, 1953 [5]. In 1954 a more detailed report of the subject was given at the Congress in Amsterdam [6]. The present report can be considered a continuation of the latter one.

We begin with the statement that analytic configurations of functions $f(z_1, \ldots, z_n)$ do not have to be complex manifolds. This is best known of the function $w = \sqrt{z_1 \cdot z_2}$; the corresponding analytic configuration has a non-spherical point for $z_1 = z_2 = 0$ which, for topological reasons, has no uniformizable neighborhood [7]. Let us consider another example: $w = \sqrt{z_1 \cdot z_2 + z_0^2}$. In the three-dimensional, analytic configuration $\mathfrak{G}$ of this function there is a family $E(t)$ consisting of two-dimensional, analytic planes which are given by the following expression:

$$E(t) = \begin{cases} w = \frac{1}{2}\left(t \cdot z_1 + \dfrac{1}{t} \cdot z_2\right) \\ z_0 = \frac{1}{2}\left(t \cdot z_1 - \dfrac{1}{t} \cdot z_2\right) \end{cases}.$$

Since

$$\begin{vmatrix} (t_1 - t_2) & \left(\dfrac{1}{t_1} - \dfrac{1}{t_2}\right) \\ (t_1 - t_2) & -\left(\dfrac{1}{t_1} - \dfrac{1}{t_2}\right) \end{vmatrix} \neq 0,$$

for $t_1 \neq t_2$, the two planes $E(t_1)$ and $E(t_2)$, $t_1 \neq t_2$ have only the point

$(w, z_1, z_2, z_0) = (0, \ldots, 0) = 0 \in \mathfrak{G}$ in common. Therefore, it is possible that two different two-dimensional planes can intersect in $0 \in \mathfrak{G}$. However, the following well-known theorem is true: Intersecting, two-dimensional, analytic planes in a three-dimensional, complex manifold always have an analytic set of dimension $d \geq 1$ in common. Therefore, $\mathfrak{G}$ cannot be a complex manifold in 0; in other words, 0 cannot be a uniformizable point of $\mathfrak{G}$.

The latter example of the analytic configuration $\mathfrak{G}$ shows a stronger pathology than $w = \sqrt{z_1 \cdot z_2}$. Although one can find an unlimited covering space of $w = \sqrt{z_1 \cdot z_2}$ consisting of uniformizable points, the same is not possible for $\mathfrak{G}$. Thus everything pointed toward a generalization of the concept of the complex manifold. This was done at about the same time by Henri Cartan [8] and by H. Behnke and K. Stein [9], however in different ways. Following this, H. Grauert and R. Remmert improved the concepts introduced by the last two authors [10]. In what follows we shall consider only the concept of complex spaces set forth by Grauert and Remmert.

The generalization of the complex manifold is achieved by the adjunction of certain singular boundary points. Since only points can be considered where the functions concerned behave algebraically, we can take our model from algebraic manifolds. The n-dimensional "algebraic manifolds" always have, in the neighborhood of a singularity, the analytic structure of a (finitely sheeted), analytically ramified covering space. Therefore we arrive at the concept of the n-dimensional, analytically ramified covering space. A triple $\mathfrak{R} = (R, \Phi, G)$ is called *an n-dimensional, finitely sheeted, analytically ramified covering space* $\mathfrak{R}$ of a domain G of the n-dimensional complex number space C^n if:

1. R is a Hausdorff space and Φ a continuous, proper, almost one-to-one mapping of R into G,†
2. there is an analytic set‡ A in G so that $\Phi^{-1}(A)$ lies nowhere dense in R and Φ is a local homeomorphism on $R - \Phi^{-1}(A)$,
3. for each point $x \in R$ there are arbitrarily small, open, connected neighborhoods U_x such that $U_x - \Phi^{-1}(A)$ is connected.

Φ is called almost one-to-one if the set $\Phi^{-1}(\mathfrak{z})$ for each $\mathfrak{z} \in G$ is discrete. "Connected" always means "arcwise connected." The points of the set $B \supset R - \Phi^{-1}(A)$ where Φ is locally topological are called schlicht points of $\mathfrak{R}$. Recently, Karl Stein has again investigated the concept of the analytically ramified covering space and achieved a further generalization.

† A continuous mapping φ of a Hausdorff-space R into a Hausdorff-space R' is called *proper*, if the inverse image of each compact subset of R' is compact in R. φ is called *almost one-to-one*, if the inverse image of each point of R' is a discrete set in R.

‡ On the concepts "analytic set" and "ordinary point of an analytic set" see [11].

He looked at ramified covering spaces of arbitrary, locally compact, locally connected topological spaces.

We continue to extend the construction of the complex space. According to property (3) of $\mathfrak{R}$, R is locally connected and $R - \Phi^{-1}(A)$ lies dense in R. $\Phi^{-1}(A)$ nowhere separates R (if U is an open, connected subset of R, then $U - \Phi^{-1}(A)$ is connected). On account of the mapping Φ, each point $x \in R$ has a base point $\mathfrak{z} = \Phi(x) \in G \subset C^n$. Since Φ maps properly, as it is almost one-to-one, there is over each point $\mathfrak{z} \in G$ only a discrete and compact set in R: thus there are only finitely many points $\Phi^{-1}(\mathfrak{z})$. It can be easily seen that over each point $\mathfrak{z} \in G - A$ there are equally many points of R (since Φ is a local homeomorphism of $R - \Phi^{-1}(A)$ onto the domain $G - A$). This number is called the *sheet number* of the covering space R of G. Since G is a domain of the number space C^n, each point of $R - \Phi^{-1}(A)$ is locally spherical, which means that in the neighborhood of such a point, R is analytically equivalent to the complex n-dimensional open sphere.

Those points of R where Φ is not a local homeomorphism are called the *ramification points* of $\mathfrak{R}$. These points are in general not locally spherical as is shown by the above examples. However, Grauert and Remmert have shown that there are only branch points over ordinary points of $A \subset G$ and these are again spherical. A ramification point $x \in R$ is called a *branch point* of $\mathfrak{R}$ if R branches in x in the same way as the function $\sqrt[k]{z}$, which means explicitly that with an appropriate numbering of the variables $z_1, \ldots, z_n$, there has to be a neighborhood $U(x)$ for $x \in R$ as well as a topological mapping ψ of U onto a domain W of the space of the complex variables $w_1, \ldots, w_n$, so that $\Phi \circ \psi^{-1}$ can be expressed as

$$\begin{aligned} z_1 &= w_1^k + h(w_2, \ldots, w_n) \\ z_2 &= w_2 \\ &\;\vdots \\ z_n &= w_n, \end{aligned}$$

where $h(w_2, \ldots, w_n)$ is a holomorphic function in W. Therefore, local uniformizing variables exist not only for schlicht points, but also for those ramification points that are branch points.

Let us consider the above mentioned facts for the particular case of the covering spaces of C^2. In this case A is complex one-dimensional, and in a neighborhood of each of its points $\mathfrak{z}_0 = (z_1^{(0)}, z_2^{(0)})$ it is determined by $f(z_1, z_2) = 0$, where f is holomorphic at $\mathfrak{z}_0$. A non-spherical point of $\mathfrak{R}$ can lie over $\mathfrak{z}_0$ only if both of the first derivatives of f vanish at $\mathfrak{z}_0$.

Non-uniformizable points are therefore isolated points in analytically ramified covering spaces of C^2.

We now come to the concept of a holomorphic mapping of an analytically ramified covering space $\mathfrak{R}$ into an analytically ramified covering space $\mathfrak{R}'$. This means a generalization of holomorphic functions on $\mathfrak{R}$.

A continuous, complex-valued function f, which is defined on an open set $W \subset R$, is called a *holomorphic function* in W, if for every schlicht point x of W there is a schlicht neighborhood $U(x) \subset W$ so that the function $f \circ (\Phi \mid U(x))^{-1}$ is holomorphic in the image $\Phi(U(x)) \subset C^n$. At the non-schlicht points it is required only that the holomorphic function be continuous. If such a point is still uniformizable, f as a function of the uniformizing variables must be holomorphic at this point. This follows from a theorem of Riemann because the exceptional points form lower dimensional sets.

Let $\mathfrak{R} = (R, \Phi, G)$ and $\mathfrak{R}' = (R', \Phi', G')$ be analytically ramified covering spaces. A continuous mapping τ of an open set $W \subset R$ into R' is called a *holomorphic mapping*, if the following is true: If $f(x')$ is an arbitrary holomorphic function in an open set W', then the function $f \circ \tau(x)$ is holomorphic in $W \cap \tau^{-1}(W')$.

The holomorphic complex-valued functions on $\mathfrak{R}$ are therefore exactly the holomorphic mappings of $\mathfrak{R}$ into the complex number plane C^1.

Now we come to the concept of complex maps on a Hausdorff space X. We shall define maps in the following way: A *map* is a pair (U, ψ) where U is an open, connected subset of X and ψ is a topological mapping of U onto an open subset of an n-dimensional, analytically ramified covering space. Two maps (U_1, ψ_1) and (U_2, ψ_2) of the Hausdorff space X are called *holomorphically compatible* if the images $\psi_1(U_1 \cap U_2)$ and $\psi_2(U_1 \cap U_2)$ of the intersection $U_1 \cap U_2$ into both of the n-dimensional, analytically ramified covering spaces are holomorphic images of each other with respect to $\psi_2 \circ \psi_1^{-1}$ or $\psi_1 \circ \psi_2^{-1}$. (They are topological images of each other by definition.) Each map has a complex dimension—namely, that of the corresponding analytically ramified covering space—and two maps for which the intersection $U_1 \cap U_2$ is not empty have the same dimension. A *complex atlas* $\mathfrak{A}$ on a Hausdorff space X is a collection of pairwise compatible maps which cover X completely.

A complex atlas $\mathfrak{A}$ on X is called *complete* if each map (U, ψ), compatible with each of the maps of $\mathfrak{A}$, belongs to $\mathfrak{A}$. Such a complete atlas on a Hausdorff space X is called a *complex analytic structure* of X. A Hausdorff space X with a complex analytic structure is called a *complex space.* Each connected component X_κ of a complex space X has a (complex) dimension $d(X_\kappa)$, namely the dimension of its maps. The dimension corresponding to X is defined by $\sup d(X_\kappa)$, where X_κ runs through all components of X.

Those fundamental concepts in order of their build-up are listed again below:

1. n-dimensional, analytically ramified covering space $\mathfrak{R}$
2. holomorphic function on $\mathfrak{R}$
3. holomorphic mappings of $\mathfrak{R}$ into $\mathfrak{R}'$
4. complete complex atlas on a Hausdorff space
5. complex space as Hausdorff space with complex analytic structure.

In general, the points of a complex space X are uniformizable. The point x in X is uniformizable if, and only if, there is a map (U^*, ψ^*) containing x, where the corresponding covering space (R, Φ, G) is the trivial covering space of the domain G — the domain G itself. This is again the case if and only if it is true for one, and consequently for every map (U, ψ) containing x, that $\psi(x)$ is a schlicht point or uniformizable ramification point of the corresponding covering space. Only such points are non-uniformizable points of the complex space X which do appear on one (and therefore on every) map as non-uniformizable ramification points. Therefore, the non-uniformizable points of a complex space form a very thin set, or more accurately, a thin set of order 2. This concept shall soon be defined in the following.

Since holomorphy of complex-valued functions is defined in each map, it is also defined in complex spaces. It is sufficient for the holomorphy of f in X if for each point $x \in X$ there is a map (U, ψ) containing x for which $f \circ \psi^{-1}$ is holomorphic in $\psi(U)$; then f is holomorphic on all maps, called holomorphic in X.

The totality of all holomorphic functions in a complex space X forms a ring $I(X)$. If X is not connected, $I(X)$ will contain divisors of zero (those functions which vanish identically on at least one connected component of X). However, if X is connected, the holomorphic functions in X form an integral ring.

The theorem of Riemann about removable singularities holds also for complex spaces. However, we cannot formulate it before we have introduced the concepts of analytic and thin sets.

Above and beyond holomorphic functions in a complex space, holomorphic mappings of complex spaces can be defined here in the same way as for the analytically ramified covering spaces.

We shall now consider the set of zeros of holomorphic functions. A subset A of a complex space X is called an *analytic set* in X if for each point $x' \in X$ there is a neighborhood U and finitely many holomorphic functions $f_1, \ldots, f_s$ in U, so that a point $x \in U$ belongs to A if and only if $f_1, \ldots, f_s$ vanish simultaneously at x. Each analytic set A of a complex space X has in each of its points x a well-defined dimension $d_x^*(A) \equiv 0(2)$ according to the dimension theory of Menger and Hurewicz. We call

$d_x = \frac{1}{2}d_x^*$ the *complex dimension* of A at x. Now we can define the thin set which already appeared in the discussion of the non-uniformizable points of complex spaces. A closed subset D of a complex space X is called a *thin set of order* k $(1 \leq k \leq d(X))$ if for each point $x \in D$ there is a neighborhood U and an analytic set A in U, so that $D \cap U \subset A, d_x(A) \leq d_x(X) - k$.

The theorem of Riemann states that in a complex space every continuous function which is holomorphic outside of a thin set D of order one is also holomorphic in D. Similarly, the theorem of Radó is valid: A function continuous in a complex space and holomorphic outside of its zeros is everywhere holomorphic in X.†

Both of these theorems can easily be proved by purely local methods. However, the theorem of Radó is much more deep-rooted than that of Riemann, as will immediately be clear for the particular case where X is the complex plane.

The definition of a meromorphic function in a complex space X can now be given in the well-known way. It is not necessary, however, to explain this in detail. We note only that the totality $K(X)$ of all meromorphic functions in a connected complex space X forms a field. In general, however, it is not possible to present every function meromorphic in X as the quotient of two functions holomorphic in X, since this is not even possible in certain domains of C^n. But we shall come to a whole class of complex spaces for which this is the case.

Little is known about analysis in neighborhoods of non-uniformizable points, but it will certainly look different from analysis in complex manifolds. Here are some examples constructed by Grauert and Remmert.

1. In general a pure $(n-1)$-dimensional analytic set in an n-dimensional complex space cannot be defined even locally as the exact zero set of a holomorphic function, although this is possible in complex manifolds as Remmert and Stein have shown [12]. It may even occur that in a neighborhood of a point this analytic set of dimension $(n-1)$ cannot be defined as a set of simultaneous zeros of less than $(n-1)$ locally holomorphic functions [13].

2. Two pure $(n-1)$-dimensional analytic sets in an n-dimensional complex space can intersect in an isolated point. But in complex manifolds such an intersection is either empty or a pure $(n-2)$-dimensional analytic set.

3. It follows from these possible exceptions for the intersection surfaces of analytic sets that there are complex spaces of arbitrarily large dimension with meromorphic functions which have isolated points of indeterminacy,

† Radó proved this theorem only for the z-plane; Behnke and Stein proved it for complex spaces. Later this proof was simplified by Cartan and Heinz [11].

while in complex manifolds these points always form analytic surfaces of dimension $(n - 2)$.

In the same way as the concrete Riemann surface (over the plane) corresponds to the abstract Riemann surface, the concrete Riemann domain corresponds to the complex space. A Hausdorff space $\mathfrak{G}$ is called a *Riemann domain*, if a mapping Φ on $\mathfrak{G}$ is given into $\bar{C}^n$, such that for each point in $\mathfrak{G}$ there are neighborhoods which are finitely-sheeted, analytically ramified covering spaces with respect to Φ. If Φ is a local homeomorphism, the Riemann domain is called *unramified*; if $\mathfrak{G}$ is connected, this would correspond to the concept of domain reported on by Behnke and Thullen in 1934. The function theory of unramified Riemann domains was developed mostly in the 1930's. This concerns the approximation theorems and the theorems of Cousin, which are analogous to the theorems of Mittal-Leffler and Weierstrass, and above all, to Oka's characterization of domains of holomorphy. (For arbitrary unramified Riemann surfaces the latter was achieved only in 1953 [14].)

A special class of complex spaces was thoroughly investigated by H. Cartan. For an introduction to the Cartan spaces we consider, next, certain covering spaces of C^n. An analytically ramified covering space (R, Φ, G) is called a *Cartan covering space* if there is a holomorphic function f on R, whose degree† over the ring of holomorphic functions on G equals to the number of sheets of the covering space. Cartan covering spaces are consequently those covering spaces which can be realized locally by analytic configurations of holomorphic functions.

We call a complex space a *Cartan space* if in the complex structure, for each point $x \in X$, there is a complex map (U, ψ) with $x \in U$, such that $\psi(U)$ is the carrier of a Cartan covering space. In Cartan spaces the fundamental theorem of Cartan and Oka is valid: If X is a Cartan space, each point $x \in X$ possesses a neighborhood $U(x)$ which can be biholomorphically mapped on a normal analytic set in a domain of the number space. An analytic set A in a domain G of C^n is called *normal* at the point $\mathfrak{z} \in A$ if each function holomorphic‡ on A at $\mathfrak{z}$ is the trace of a function which is holomorphic on an open neighborhood $V(\mathfrak{z}) \subset G$. The set A is called normal if it consists of normal points only.

Until now, function theory in its basic sense has been developed for Cartan spaces only. The holomorphically complete complex spaces

† Each element $f \in I(R)$ annuls an irreducible pseudo-polynomial with coefficients holomorphic on G. The degree of f is the degree of this pseudo-polynomial. The degree of $I(R)$ over $I(G)$ is defined as the maximum of the degrees of the elements $f \in I(R)$ over $I(G)$.

‡ Here the following concept is used: "holomorphic on an analytic subset." For this, the function need not be defined in a whole neighborhood of the analytic set [15].

which shall be introduced in Section 4 are in particular such Cartan spaces. It is still unknown but nevertheless widely discussed whether or not every complex space is a Cartan space.§

2. Holomorphic mappings and analytic decompositions

1. Holomorphic mappings of complex spaces have been explained as a generalization of holomorphic functions in Section 1. In the following we shall investigate holomorphic mappings in detail.

If $\tau : X \to Y$ is an arbitrary holomorphic mapping and B an analytic set in Y, then $\tau^{-1}(B)$ is always an analytic set in X. The inverse images of the points of Y are called the *fibres* of the mapping τ. The complex codimension at the point x of the fibre $\tau^{-1}(\tau(x))$, $x \in X$ is called the *rank* of τ at x, with the symbol $r_x(\tau)$. The number $r(\tau) = \sup_{x \in X} r_x(\tau)$ is called the rank of the mapping τ. The set E of all points $x \in X$ with $r_x(\tau) < r(\tau)$ is called the *set of degeneration* of the mapping τ. If E is empty, τ is said to be nowhere degenerated. It can be shown that:

THEOREM 1. The set of degeneration E of a holomorphic mapping $\tau : X \to Y$ is always an analytic set in X [17].

If $\tau : X \to Y$ is a holomorphic mapping, then the set $\tau(X)$ is not in general an analytic set. It is true, however, that:

THEOREM 2. If $\tau : X \to Y$ is a nowhere degenerated mapping of rank r, then each point $x \in X$ possesses arbitrarily small neighborhoods $U(x)$ such that $\tau(U(x))$ is a pure r-dimensional analytic set in a neighborhood $V(\tau(x)) \subset Y$.

With the application of these results one can show:

THEOREM 3. A holomorphic mapping $\tau : X \to Y$ of rank r into an r-dimensional complex space Y is nowhere degenerated if and only if it is open.

τ is called open if the images of open sets are open.

The following theorem is of basic importance for the mapping theory.

THEOREM 4. If $\tau : X \to Y$ is a proper holomorphic mapping of rank r, then $\tau(X)$ is an irreducible r-dimensional analytic set in Y.

Here a mapping is called *proper* if the inverse images of compact sets are always compact. The given theorem is false if the mapping τ is not assumed to be proper.

Theorem 4 can be extended to so-called meromorphic mappings. In

§ Serre has defined complex spaces more generally than we have, such that according to his concept each analytic subset of a complex space is again a complex space. After him the Cartan spaces mentioned in our paper are called "normal spaces" [16]. Recently Grauert and Remmert succeeded in proving that every complex space in our sense is a normal complex space in the sense of Cartan and Serre.

so doing, it is possible to derive important theorems about analytic and algebraic dependence of meromorphic functions in compact complex spaces [18], [19].

Holomorphic mappings can be characterized by their graphs:

THEOREM 5. A mapping $\tau : X \to Y$ of a complex space X into a complex space Y is holomorphic if and only if its graph—that is, the set of all points $(x, \tau(x))$ in the product space $X \times Y$—is an analytic set in $X \times Y$ of dimension $d(X)$. The graph of τ is then a closed complex subspace of $X \times Y$ [20].

An analytic set T in a complex space X is called a *complex subspace* of X if it is possible to introduce a complex structure in T, which is compatible with the induced topology, so that the injection $\iota : T \to X$ is a holomorphic mapping. It can be shown [21] that the complex subspaces of a complex space X are exactly those locally irreducible analytic sets in X provided with the natural complex structure.

It follows from Theorem 5 that a one-to-one holomorphic mapping $\tau : X \to Y$ of a complex space X into a space Y of the same dimension is always biholomorphic.

2. Each holomorphic mapping $\tau : X \to Y$ defines a decomposition of the space X; the elements of this decomposition are the fibres of the mapping τ. We now consider generally the case of decompositions of a complex space X, the elements of which are analytic sets in X. Such a decomposition Z will be called analytic if there is a complex structure $\mathfrak{A}$ on the quotient set X/Z so that the natural projection $\varphi : X \to X/Z$ is holomorphic, whose value at the point $x \in X$ is the unique element of Z to which x belongs. Each such complex structure $\mathfrak{A}$ is called a complex structure corresponding to the decomposition Z; the set X/Z with this structure is called a complex space corresponding to Z. Let the decomposition Z be defined by a holomorphic mapping $\tau : X \to Y$, in symbols $Z = Z(\tau)$. Hence $Z(\tau)$ will surely be an analytic decomposition of X if τ maps the space X onto Y.

If Z is an analytic decomposition of X, the complex structure on the quotient set X/Z is not uniquely defined by Z. One can, however, show [22]:

THEOREM 6. If X is a complex space with countable topology, and Z an analytic decomposition of X, there is, at most, one complex structure on each topological decomposition space X/Z conjugate to Z.

As explanation, it is possible to introduce topologies in the quotient set X/Z in various ways, so that the natural mapping $\iota : X \to X/Z$ remains continuous. If one added any such topologies to X/Z, the resulting topological spaces are called *topological decomposition spaces* conjugate to the decomposition Z. Furthermore:

THEOREM 7. If X has a countable base, and if there are two decomposition spaces conjugate to the analytic decomposition Z of X and which have non-equivalent but comparable topologies, then, at most, one of these spaces can carry a complex structure corresponding to Z.

Especially interesting are such analytic decompositions Z of X for which the quotient space X/Z (that is, the decomposition space conjugate to Z with the finest topology) has a complex structure. An analytic decomposition Z of X is called *normal* if there is one and only one complex structure corresponding to Z on the quotient space X/Z. The quotient space provided with this complex structure is called the *complex decomposition space* corresponding to Z.

In order to be able to form a criterion for the case when an analytic decomposition is normal, we introduce the concept of proper decomposition. A decomposition Z of X is called *proper* if the saturated envelope of each compact set K of X is compact. (The saturated envelope of K means the set of all of those points of X which are Z-equivalent with at least one point of K.)

Now we can state [23]:

THEOREM 8. Every proper analytic decomposition Z of a complex space X is normal. Besides the complex quotient space X/Z corresponding to Z, there is no further complex space corresponding to Z.

3. A decomposition of a complex space X is called *simple* if its elements are all connected. Each decomposition Z of X gives rise to a simple decomposition Z' of X; the elements of Z' are the connected components of the elements of Z. If Z (respectively Z') is an analytic decomposition of X, then Z' (respectively Z) is not necessarily an analytic decomposition of X.

The following theorem gives a sufficient condition that the simple decomposition defined by a holomorphic mapping of the complex space be analytic [24].

THEOREM 9. Let X be a complex manifold and $\tau : X \to Y$ a holomorphic mapping of X into a complex space Y in such a fashion that the connected components of the fibres of τ are all compact. Then the simple decomposition $Z'(\tau)$ of X defined by τ is a normal analytic decomposition. The natural mapping of X onto the complex decomposition space $X/Z'(\tau)$ is a proper holomorphic mapping.

The conclusion of this theorem remains correct even if X is an n-dimensional Cartan space and τ a holomorphic mapping, the rank of which is larger than $n - 2$. Until now, Theorem 9 has not been proved for mappings of the rank $\leq n - 2$ for Cartan spaces.

4. Let X, X_1, X_2 be complex spaces and $\tau_1 : X \to X_1$, $\tau_2 : X \to X_2$ be holomorphic mappings. The mapping τ_2 is called *dependent* on τ_1 if the decomposition $Z'(\tau_1)$ is finer than the decomposition $Z'(\tau_2)$. (A decomposition Z_1 is called *finer* than a decomposition Z_2 if each element of Z_1 is contained in an element of Z_2.) This is the case if and only if τ_2 is constant

on each connected component of the fibres of τ_1. This concept of dependence clearly corresponds with the known concept of analytic dependence of holomorphic functions in domains G of C^n ($X = G$, $X_1 = X_2 = C^1$).

Two holomorphic mappings $\tau_1 : X \to X_1$, $\tau_2 : X \to X_2$ are called *related* if τ_1 is dependent on τ_2 and τ_2 is dependent on τ_1; this is true if and only if $Z'(\tau_1) = Z'(\tau_2)$.

Given a holomorphic mapping $\tau_1 : X \to X_1$, a pair (Φ, X^*) is called a *complex base* corresponding to τ_1 if the following is true:

(a) X^* is a complex space; $\Phi : X \to X^*$ is a holomorphic mapping of X onto X^* related to τ_1.

(b) If $\tau_2 : X \to X_2$ is an arbitrary holomorphic mapping dependent on τ_1, there is a holomorphic mapping $\tau : X^* \to X_2$ so that $\tau_2 = \tau \circ \Phi$.

The space X^* is called a *complex base space* for the mapping τ_1. If there exists a complex base (Φ, X^*) for the mapping $\tau_1 : X \to X_1$, investigation of the classes of holomorphic mappings dependent on τ_1 is equivalent to investigation of the corresponding holomorphic mappings of X^*.

It can then be shown:

The complex base space for a holomorphic mapping τ_1 (in case it exists) is uniquely determined, except for analytic homeomorphisms.

It can be proved further:

It is necessary for the existence of a complex base for a holomorphic mapping $\tau_1 : X \to X_1$ that there be a finest analytic decomposition $\tilde{Z}$ between the decompositions $Z'(\tau_1)$ and $Z(\tau_1)$.

From Theorem 9 follows [25]:

THEOREM 10. Let $\tau_1 : X \to X_1$ be a holomorphic mapping of a complex manifold X into a complex space X_1 so that the connected components of the fibres of τ_1 are compact. Then there exists a complex base (Φ, X^*) corresponding to τ_1; the complex base space X^* is the complex decomposition space $X/Z'(\tau_1)$ corresponding to $Z'(\tau_1)$, and Φ is the natural mapping of X onto $X/Z'(\tau_1)$.

Further applications of Theorem 9 refer to reduction problems of complex spaces [26].

The assumption of Theorem 10, that the connected components of the fibres of the holomorphic mapping in question are compact, is essential. It is, however, possible to define analytic decompositions $\tilde{Z}$ for mappings τ which do not satisfy this condition, such that $\tilde{Z}$ lies between $Z'(\tau_1)$ and $Z(\tau_1)$ [27, 28].

3. Modifications

1. While in the case of classical function theory we have only one possibility for closing the complex affine space C^1, in the case of functions of n

complex variables there are infinitely many such possibilities. This led to the concept of modification [29].

Attempts had been made to solve the following basic problem: Let C^n be closed in a certain way; given a function in this space, what happens to the behavior of the function if the closure is changed? Later it was noticed that in algebraic geometry, modification was well known as the important process of resolution of singularities (non-uniformizable ramification points). We should note that the above-mentioned problem in algebraic geometry was solved only for dimensions 2 and 3 [30].

Fritz Hirzebruch has used modifications to complete analytic configurations A of functions $f(z_1, z_2)$ [31]. (Here as well as in classical theory the non-essential singularities are also added to the analytic configuration.) The non-uniformizable points and the points of indeterminacy are replaced in A. The unchanged part is a complex manifold A^*. This manifold is continued to a complex manifold $\tilde{A}$ so that each of the earlier exceptional points is replaced by a union of finitely-many, compact, irreducible analytic surfaces. f is now a meromorphic function on $\tilde{A}$ without points of indeterminacy.

In the meantime, papers of Grauert, Kreyszig, Remmert, and Stoll have extended the theory of modifications far beyond this first goal [32]. In describing this theory we shall limit our discussion to connected complex spaces X, $'X$ which are both of the same dimension.

Let N be a thin set of order 1 in X, $'N \neq {}'X$ be closed in $'X$. A quintuple $('X, 'N, \tau, N, X)$ is called a *modification of X in N* if the following conditions are satisfied:

1. τ is a biholomorphic mapping of $'X - {}'N$ onto $X - N$.
2. If U is any neighborhood of N, then $\tau^{-1}(U - N) \cup {}'N$ is a neighborhood of $'N$.

The modification $('X, 'N, \tau, N, X)$ is called *continuous* if, instead of 2, the following more extensive condition is satisfied:

2′. Let $\{U_j : j \in J\}$ be any open covering of N; then $\bigcup_{j \in J} V_j$, $V_j =$ open kernel of $\tau^{-1}(U_j - U_j \cap N) \cup {}'N$ is a neighborhood of $'N$. In the following we shall limit our discussion to continuous modifications; more general types of modifications were extensively considered, for example, by W. Stoll. The following necessary and sufficient criterion holds: A modification $('X, 'N, \tau, N, X)$ is continuous if and only if the mapping $\tau : {}'X - {}'N \to X - N$ can be continued to a holomorphic mapping $\tau : {}'X \to X$ of $'X$ into X. The set $'N$ then is a thin set of order 1 in $'X$.

The proof of this theorem is based mainly on a theorem of Radó. Part of the conclusion of this theorem had already been proved in the first report on modifications. The above form of the theorem is due to Grauert and Remmert. From this theorem it follows in a trivial manner that there is a relation between the meromorphic (similarly the holomorphic) functions in X and $'X$. Let $K(X)$ (let $I(X)$) be the field (be the integral ring) of meromorphic (of holomorphic) functions in X. Then we have:

(a) If $('X, 'N, \tau, N, X)$ is a continuous modification, then there is a natural monomorphism $^*\tau : K(X) \to K('X)$ which maps $I(X)$ into $I('X)$.

Let $V(X)$ be the lattice of analytic sets in X. Then we have:

(b) If $('X, 'N, \tau, N, X)$ is a continuous modification, and if the continued mapping τ maps the space $'X$ onto X, there is a natural monomorphism $\hat{\tau} : V(X) \to V('X)$.

We should note that $^*\tau$ and $\hat{\tau}$ are not necessarily isomorphisms. Counterexamples are given by H. Grauert and R. Remmert.

2. In the following we shall consider continuous modifications exclusively. Let τ denote generally the mapping of $'X$ into X, which is determined by the continuation of $\tau : 'X - 'N \to X - N$. We limit ourselves further to genuine modifications. A continuous modification $('X, 'N, \tau, N, X)$ is called *genuine* if N is non-empty and if each fibre $\tau^{-1}(x)$, $x \in N$ contains at least two points or no points at all.

In order to justify the restriction to genuine modifications we shall now introduce the concept of equivalence of modifications in a natural manner.

Two modifications $('X, 'N_i, \tau_i, N_i, X)$, $i = 1, 2$ are called *equivalent* if $\tau_1 \equiv \tau_2$. Furthermore, a modification $('X, 'N, \tau, N, X)$ is called the *identity* if τ is a one-to-one mapping of $'X$ onto X.

Then this important theorem is true:

For every continuous modification $('X, 'N, \tau, N, X)$ that is not the identity, there exists exactly one equivalent genuine modification $('X, '\dot{N}, \tau, \dot{N}, X)$. $'\dot{N}$ is the set of degeneration of τ in $'X$; hence $'\dot{N}$ is either empty or an analytic set in $'X$ without isolated points. $\dot{N}$ is non-empty.

Grauert and Remmert's proof is based on general theorems from the theory of holomorphic mappings. We note that $\dot{N}$ is not necessarily an analytic set in X, not even in the case where $\tau('X) = X$.

It follows immediately from the above-mentioned reduction theorem that a genuine modification between two Riemann surfaces $'X$, X always consists of eliminating from X a non-empty closed set $\dot{N}$.

A very precise statement can be made about the set $'\dot{N}$ in a genuine modification $('X, '\dot{N}, \tau, \dot{N}, X)$ if X is a complex manifold. In this case the following holds:

If $('X, '\dot{N}, \tau, \dot{N}, X)$ is a genuine modification of an n-dimensional complex manifold X, the set $'\dot{N}$ is empty or pure $(n-1)$-dimensional.

Similarly, this theorem is a special case of a general theorem on sets of degeneration of holomorphic mappings. However, the above theorem loses its validity if X has non-uniformizable points.

3. Proper modifications are the most important ones. A continuous modification $('X, 'N, \tau, N, X)$ is called *proper* if $\tau : 'X \to X$ is a proper mapping. (We repeat, a mapping is called proper if the inverse image of each compact set is compact.)

In the case of a proper modification $('X, 'N, \tau, N, X)$, the mapping τ is always a mapping of $'X$ onto X. The following simple criterion holds:

A modification $('X, 'N, \tau, N, X)$ is proper if and only if $\tau('X) = X$ and all fibres of τ are compact. From the theory of proper holomorphic mappings we immediately obtain the following result. If $('X, '\dot{N}, \tau, \dot{N}, X)$ is a genuine proper modification of an n-dimensional complex space, then $'\dot{N}$ and $\dot{N}$ are analytic sets. $\dot{N}$ is at most $(n-2)$-dimensional.

If a modification is continuous then there exists a monomorphism between the holomorphic functions in X and those in $'X$, and a similar statement is true with respect to meromorphic functions and analytic sets in X and in $'X$. If a modification is in addition also proper, then the following statement holds:

(a) Let $('X, 'N, \tau, N, X)$ be a proper modification. Then the natural monomorphism $^*\tau : K(X) \to K('X)$ is an isomorphism and $^*\tau(I(X)) = I('X)$.

(b) Let $('X, 'N, \tau, N, X)$ be a proper modification. Then the inverse $\hat{\tau}^{-1}$ of the natural monomorphism $\hat{\tau} : V(X) \to V('X)$ can be continued to an epimorphism $\tilde{\tau}^{-1} : V('X) \to V(X)$.

4. Two complex spaces $'X$, X will be called *related* if there exist finitely many complex spaces $X = X_0, X_1, \dots, X_r = 'X$ with the following property:

X_ρ can be obtained from $X_{\rho-1}$ by a proper modification, or vice versa; here $0 < \rho \leq r$. Then we have the following immediate result:

If $'X$ and X are related complex spaces, $K('X)$ and $K(X)$ as well as $I('X)$ and $I(X)$ are naturally isomorphic.

Examples of related complex spaces are the multiple-projective spaces of the same dimension.

Two complex spaces $'X$ and X are related if and only if there is a meromorphic mapping $\mu : X \to 'X$ of X onto $'X$. Here a mapping $\mu : X \to 'X$ of a space X into a space $'X$ is called meromorphic if it has the following properties:

(a) To each point $x \in X$ there corresponds a non-empty compact set $\mu(x) \subset 'X$.

(b) The graph L_μ of μ (i.e. the totality of all points $(x, 'x) \in X \times 'X$ where $'x \in \mu(x)$) is an analytic subset of $X \times 'X$ which has the same dimension as X.

(c) There is a dense set $*X$ in X such that $\mu(x)$, $x \in *X$ consists of one point.

A meromorphic mapping $\mu : X \rightarrow Y$ gives rise to a proper modification of X in a natural way. Namely, there is an analytic set N in X and an analytic set $'N$ in the graph L_μ of μ such that $(L_\mu, 'N, \pi, N, X)$ is a proper modification. Here, π denotes the projection of L_μ onto X.

5. Complex spaces which cannot be obtained from other spaces by genuine proper modifications are called *complex primitive spaces.* Obviously, each complex space $'X$ is primitive if each compact analytic set in $'X$ is zero-dimensional or is identical with $'X$. In particular the K-complete spaces [33] and hence the holomorphically complete spaces [34] are complex primitive spaces. However, to show the existence of compact, complex primitive spaces (note: each K-complete space is non-compact) we introduce the following terminology:

A pure k-dimensional analytic set A of a complex space X is said to be *isolated* in X if there is a neighborhood $U(A) \subset X$ such that each analytic set which is pure k-dimensional in $U(A)$ is contained in A. It follows that if $('X, '\dot{N}, \tau, \dot{N}, X)$ is a genuine proper modification, and if X has a countable topology, then each pure-dimensional component of $'\dot{N}$ is isolated in $'X$. An immediate consequence of this theorem is:

Every homogeneous, compact, complex manifold is a primitive space.

Thereby, a complex space is called homogeneous if its group of analytic automorphisms operates on the space in a transitive way.

6. Proper modifications can be regarded as special proper analytic decompositions. As explanation:

If $('X, 'N, \tau, N, X)$ is a proper modification, then the analytic decomposition $Z(\tau)$ generated by τ is simple and proper. Its rank equals the dimension of $'X$.

The converse of this assertion is also valid:

If Z is an arbitrary, simple, proper analytic decomposition of a complex space of $'X$ of the rank dim $('X)$, then there exist an analytic set $'N$ in $'X$ and an analytic set N in the quotient space $'X/Z$, such that $('X, 'N, \pi, N, 'X/Z)$ is a proper modification. Here, π is the natural mapping of $'X$ onto $'X/Z$.

7. Perhaps the most important example of a proper modification is the generalized Hopf σ-process. This process consists of the replacement of a k-dimensional complex submanifold N of an n-dimensional complex manifold X by a complex fibre space $'N$ with the base N and the projective space P^{n-k-1} as fibre. The new space $'X$ is again a complex manifold.

A proper modification is called a *σ-modification* if it can be obtained by iterated σ-processes. A conjecture, which has been proved only for the case of dimension 2, asserts: every proper modification between complex manifolds is a σ-modification.

4. Function theory on holomorphically complete spaces

A. Function theory in one-dimensional complex manifolds is generally known to be the theory of holomorphic and meromorphic functions on Riemann surfaces. Nevertheless, we shall briefly formulate some of the interesting fundamental theorems, but in a terminology in which they can be carried over (though perhaps not unconditionally) to higher dimensions.

Let R be an arbitrary, non-compact Riemann surface. In the following discussion of holomorphic or meromorphic functions on R it is assumed that they are unique. Then these assertions hold [35]:

1. Separation: If x_1 and x_2 are different points of R, there is always a holomorphic function f on R such that $f(x_1) \neq f(x_2)$.
2. Existence of holomorphic, local uniformizing variables: For each point $x \in R$ there is a holomorphic function t on R which uniformizes a neighborhood of x.
3. Holomorphic convexity: For each relatively compact set B the set of points $x \in R$, for which all holomorphic functions f on R satisfy $|f(x)| \leq \sup |f(B)|$, is compact in R.
4. The generalized Mittag-Leffler Theorem (Cousin I): Let there be assigned to each point $x \in R$ a neighborhood $U(x) \subset R$ and a meromorphic function f_x in $U(x)$. Let $f_x - f_y$ be holomorphic in $U(x) \cap U(y)$. Then there is a meromorphic function F on R so that $F - f_x$ is holomorphic in $U(x)$.
5. The generalized Weierstrass Theorem (Cousin II): Let there be assigned to each point $x \in R$ a neighborhood $U(x) \subset R$ and a function f_x holomorphic in $U(x)$. Let f_x/f_y be holomorphic in $U(x) \cap U(y)$ and different from zero. Then there is a holomorphic function F on R such that F/f_x is holomorphic and different from zero in $U(x)$.
6. Poincaré's Theorem on R: If $F(x)$ is a meromorphic function on R, then there are two holomorphic functions $g_1(x)$, $g_2(x)$ on R which are locally relatively prime such that $F(x) = g_1(x)/g_2(x)$ on R.

These theorems can easily be proved with the approximation theorem (a generalization of Runge's theorem) [36]. Let f be holomorphic in G; let G be simply connected relative to a larger domain $^*G$ containing G (this means that each cycle of G bounding in $^*G$ also bounds in G). Then f in G is the limit-function of a sequence of functions f_n, uniformly convergent in the interior of G, where each f_n is holomorphic in $^*G$.

B. Certainly not all complex manifolds of higher dimension have the foregoing six properties. E. Calabi and B. Eckmann [37] have shown that there are non-compact complex manifolds M^n even of the topological type of $2n$-dimensional cells, where the ring of holomorphic functions on M^n consists merely of constants. Obviously, the above-mentioned properties 1 through 5 cannot be satisfied on such manifolds. While for $n = 1$ the cell can have only two complex structures (that of the circle and that of the open plane), the $2n$-dimensional cell for $n > 1$ has a set of complex structures which has the power of the continuum, among them the "function-scarce" structures discovered by the above-mentioned authors. In order to carry function theory onto M^n, special conditions have to be satisfied on this manifold.

Therefore, Karl Stein has introduced the concept of the holomorphically complete (complex) manifold (variété de Stein) [38]. A complex manifold M is called a *holomorphically complete* manifold if the properties 1–3 of A are fulfilled, and if M has a countable base. The property (2) is to be carried over to several variables in the following way. For each point $x \in M$ there must be n holomorphic functions on $M = M^n$ which uniquely map a neighborhood $U(x)$ onto a domain in C^n. The additional axiom of countable topology seemed to be unavoidable at this stage, for Heinz Hopf, L. Calabi, and M. Rosenlicht had given examples of complex manifolds which do not have a countable topology.

These holomorphically complete manifolds essentially show those function-theoretical properties which we have known on Riemann surfaces. The following approximation theorem holds [39]:

If $^*M^n$ is a holomorphically complete manifold which is a subdomain of a holomorphically complete manifold M^n, and if $^*M^n$ can be extended continuously over holomorphically complete manifolds to M^n, then every holomorphic function in $^*M^n$ can be approximated in the interior of $^*M^n$ by functions holomorphic in M^n.

Furthermore, the first theorem of Cousin is valid [40]: for arbitrarily given principal parts in M—or what means the same: for each given distribution of local meromorphic functions satisfying the compatibility conditions—there is a meromorphic function in M with the same poles. Cousin II does not always hold. In 1917 R. H. Gronwall [41] proved for a special case that topological conditions have to be fulfilled for the existence of holomorphic functions f with preassigned zeros. After Oka had proved in 1937 [42] that the function exists in domains of holomorphy if and only if there is a continuous function with the preassigned zeros, K. Stein was able to give partly necessary and partly sufficient conditions for the existence of a function f for preassigned surfaces of zeros in a holomorphically complete manifold [43].

Furthermore, in 1953 H. Cartan and J. P. Serre have listed necessary and sufficient cohomology conditions. Where Cousin II is valid, Poincaré's assertion concerning the representation of meromorphic functions as quotients of holomorphic functions is true.

C. In 1950 Leray founded the theory of sheaves which was soon applied successfully in topology. Today one defines a sheaf (of Abelian groups) over a topological space X as a triple $\mathfrak{S} = (S, \pi, X)$ with the following properties [44]:

1. S is a topological space, π maps S onto X and is a local homeomorphism.
2. Each stalk $S_x = \pi^{-1}(x)$, $x \in X$, possesses a binary operation, such that S_x is an Abelian group.
3. The mapping $(S'_x, S''_x) \to S'_x - S''_x$ of the space $\bigcup_{x \in X} S_x \times S_x$ onto S is continuous with respect to the relative topology, which is induced by the product topology of $S \times S$.

The most important examples of sheaves over X are the sheaves of germs of (continuous, etc.) functions. If X is a complex manifold (or a complex space) the sheaf $\mathcal{O}^q$ of germs of holomorphic mappings in C^q, $q = 1, 2, 3, \ldots$, is especially interesting. H. Cartan and J. P. Serre among others have used the sheaf $\mathcal{O}^1$ to define the concept of analytic sheaf in order to establish the premises necessary for the application of sheaf theory to function theory of several complex variables.

A sheaf $\mathfrak{S} = (S, \pi, X)$ is an *analytic sheaf* if:

1. the ring $\mathcal{O}^1_x$ operates on S_x.
2. the mapping $\bigcup_{x \in X} \mathcal{O}^1_x \times S_x \to S$ thereby defined is continuous.

In particular $\mathcal{O}^q$ is an analytic sheaf. Homomorphisms and sequences are important aids in sheaf theory. A homomorphism $\varphi : \mathfrak{S}' \to \mathfrak{S}$ of an analytic sheaf $\mathfrak{S}' = (S', \pi', X)$ into an analytic sheaf $\mathfrak{S} = (S, \pi, X)$ is a continuous mapping of S' into S which maps each group S'_x homomorphically over $\mathcal{O}_x$ into S_x. A sequence of homomorphisms $\mathfrak{S}_1 \xrightarrow{\varphi_1} \mathfrak{S}_2 \xrightarrow{\varphi_2} \mathfrak{S}_3$ is called *exact* at $\mathfrak{S}_2$ when the image $\varphi_1(S_1)$ is equal to the kernel of φ_2. An arbitrary sequence $\mathfrak{S}_1 \to \ldots \to \mathfrak{S}_k$ is called exact if it is everywhere exact.

Sheaf theory takes on special meaning because of the fact that cohomology groups $H^\nu(X, \mathfrak{S})$, $\nu = 0, 1, \ldots$, with coefficients in the sheaf $\mathfrak{S}$ can be defined. The group $H^0(X, \mathfrak{S})$ coincides with the group of sections in $\mathfrak{S}$ over X (A *section* in $\mathfrak{S}$ over X is a continuous mapping s of X into $\mathfrak{S}$ with $\pi \circ s(x) = x$ for $x \in X$). If the "zero"-sheaf is denoted by O, it follows that to each exact sheaf sequence $O \to \mathfrak{S}_1 \xrightarrow{i} \mathfrak{S}_2 \xrightarrow{j} \mathfrak{S}_3 \to O$ there corresponds an exact sequence $O \to H^0(X, \mathfrak{S}_1) \xrightarrow{i} H^0(X, \mathfrak{S}_2) \xrightarrow{j} H^0(X, \mathfrak{S}_3) \to H^1(X, \mathfrak{S}_1) \xrightarrow{i} H^1(X, \mathfrak{S}_2) \xrightarrow{j} \ldots$ of the cohomology groups.

Of greater interest in complex analysis are the coherent analytic sheafs. We define with Serre: An analytic sheaf $\mathfrak{S}$ over a complex manifold X is called *coherent* if one can define locally an exact sequence $\mathcal{O}^p \to \mathcal{O}^q \to \mathfrak{S} \to O$ everywhere in X. If A is an analytic subset of X and if $\mathcal{O}_A$ is the sheaf of germs of holomorphic functions which vanish on A, then in particular $\mathcal{O}_A$ has the property of coherence as shown by H. Cartan.

H. Cartan was able to deduce two fundamental theorems for coherent sheaves $\mathfrak{S}$ over holomorphically complete manifolds, which include the statements Cousin I and Cousin II as well as the Poincaré Theorem [45].

THEOREM A: If $S'_x = S_x \cap \{s\}$, where $\{s\}$ designates the set of sections $H^0(X, \mathfrak{S})$, then S'_x generates the group S_x over $\mathcal{O}_x$.

THEOREM B: $H^\nu(X, \mathfrak{S}) = O$ for $\nu = 1, 2, \ldots$.

It follows from Theorem A for $\mathfrak{S} = \mathcal{O}_A$ that the analytic set $A \subset X$ is exactly the set of simultaneous zeros of all holomorphic functions vanishing on A. One of the authors deduced from this for an n-dimensional homomorphically complete manifold X that there are always $n + 1$ holomorphic functions in X which vanish simultaneously exactly on A. A further simple deduction from Theorem A is that each meromorphic function in X is the quotient of two holomorphic functions (which of course do not have to be locally relatively prime).

Let $\mathcal{M}$ be the sheaf of germs of meromorphic functions; then $\mathcal{O}^1$ is a subsheaf of $\mathcal{M}$. If one forms the quotient group $\mathcal{M}_x/\mathcal{O}^1_x$ for each $x \in X$, one obtains a sheaf $\mathcal{H}$, the points of which are germs of principal parts of meromorphic functions. A section s in $\mathcal{H}$ corresponds to a preassigned Cousin-I-distribution. If i is the identical homomorphism $\mathcal{O}^1 \to \mathcal{M}$, and j the quotient homomorphism $\mathcal{M} \to \mathcal{H}$, one has the exact sequence $O \to \mathcal{O}^1 \xrightarrow{i} \mathcal{M} \xrightarrow{j} \mathcal{H} \to O$ and the corresponding sequence of cohomology $O \to H^0(X, \mathcal{O}^1) \xrightarrow{i} H^0(X, \mathcal{M}) \xrightarrow{j} H^0(X, \mathcal{H}) \to O$, since according to Theorem B, $H^1(X, \mathcal{O}^1) = O$. That, however, means nothing else than that each section in $\mathcal{H}$ is the image by j of a section in $\mathcal{M}$, therefore of a meromorphic function f: For preassigned principal parts there always exists a meromorphic function. In a similar way, using an exact sequence, one can derive the theorem Cousin II. This, however, cannot be given in the present report.

D. Thus we have described the state of research at the time of the Brussels colloquium on function theory of several complex variables in March 1953 [46]. One can describe this state by analogy with classical function theory: classical function theory was at the same stage between the appearance of H. Weyl's classical book and the paper of T. Radó in 1922 [47]. At that time there existed a system of axioms for the abstract Riemann surface—that given by Hermann Weyl. The axiom of triangulation was added to the other axioms generally assumed today. T. Radó showed with the uniformization theorem that the existence of a countable base for open sets of an abstract Riemann surface R follows from the other axioms without assuming triangulation. It follows from this that R can be triangulated. Therefore this property is no longer required in any modern axiomatic development of Riemann surfaces.

Even before the colloquium in Brussels, investigations of H. Hopf, L. Calabi, and M. Rosenlicht [48] had shown that an analogous theorem for complex manifolds of dimension higher than one does not hold. The question had remained open for holomorphically complete manifolds. Then one of the authors [49], gave the proof that the assumption of countable topology is unnecessary for holomorphically complete manifolds, which are in this respect analogous to non-compact Riemann surfaces.

Thus the concept of the holomorphically complete manifold can finally be developed. According to new definitions, a holomorphically complete manifold M is a complex manifold satisfying the following two axioms:

1. M is K-complete, or in more exact terms—for each point $x \in M$ there are finitely many holomorphic functions $f_1, \ldots, f_k$ on M which provide an almost one-to-one mapping τ of a neighborhood $U(x)$ into C^k. Almost one-to-one means that the set $\tau^{-1}(\mathfrak{z}) \cap U$ for each $\mathfrak{z} \in C^k$ always consists of isolated points.

2. M is holomorphically-convex.

While all persons interested in holomorphical-convexity have necessarily become very familiar with this concept during the last twenty years, K-completeness is a concept that has been newly introduced. Consider, for instance, K-completeness in the case of a concrete Riemann domain $\mathfrak{G}$ over C^n, which is mapped almost one-to-one into C^n by the natural projection on its base points and hence K-complete.

Certainly in Axiom 1 the number k of the mapping functions is not smaller than n, since otherwise, whole analytic surfaces would be projected by τ onto points. At this moment we shall not discuss how large the number k will have to be chosen. It will be shown later that k can always be chosen equal to n. The above axioms are weaker than those given by K. Stein. However, they determine the same class of manifolds. This will be shown as follows:

The existence of a countable base for open sets of M follows from K-completeness. In the proof of this theorem it is first shown that the space $\mathfrak{H}$ of holomorphic functions on M whose topological structure is defined by uniform convergence in the interior of M is separable. It follows that there is a countable dense subset $\mathfrak{F}$ of $\mathfrak{H}$. This statement was proved with S. Bergman's theory of orthogonal functions by giving a countable orthogonal system in a subdomain $M' \Subset M$ for the functions of $\mathfrak{H}$. Each holomorphic function of $\mathfrak{H}$ can be approximated by functions of $\mathfrak{F}$, and, making use of the K-completeness of M, it follows that to each point of M there are finitely many functions of $\mathfrak{F}$ which provide an almost one-to-one mapping of a neighborhood of this point. Assuming that M has an uncountable topology it follows further from the countability of $\mathfrak{F}$ that there is a connected subspace M_0 of M without countable topology which is mapped almost one-to-one by finitely many functions $f_1, \ldots, f_k \in \mathfrak{F}$ into C^k. M_0 is mapped by $f_1, \ldots, f_k$ one-to-one onto a (perhaps lower dimensional) Riemann domain over the C^k, which itself naturally possesses a countable base. Contrary to our assumption, M_0 would have a countable base as well.

The other requirements of K. Stein, the separation axiom and the existence of local coordinates which can be defined by functions holomorphic in all of M, do not follow from K-completeness exclusively, in contrast to the case of countable topology. For, there exist ramified Riemann domains $\mathfrak{G}$ over the C^n, whose envelopes of holomorphy are schlicht. If x and y are two points of $\mathfrak{G}$ over the same base point, then each holomorphic function in $\mathfrak{G}$ has the same function-element in x and y and therefore the same function value. Hence the separation axiom does not hold in $\mathfrak{G}$. K-completeness alone does not establish the existence of local uniformizing variables defined by holomorphic functions on $\mathfrak{G}$. For, if x_0 is a ramification point of $\mathfrak{G}$, then in every neighborhood of x_0 there are different points with the same base point, which are mapped onto the same point of the complex number space by each holomorphic mapping of $\mathfrak{G}$.

If it is required that M be holomorphically convex, it can be shown that M has the two above-mentioned properties. For each point $x \in M$ there exists a certain neighborhood U which is a holomorphically complete manifold in the sense of Stein. Now with Oka's successfully applied linking method, we can link suitable neighborhoods U and show that the result of this linking is again a holomorphically complete manifold (in the sense of Stein). Thus one can show by finite repetition of this process that each analytic polyhedron of M satisfies the Stein conditions 1,2. Because of holomorphic-convexity this also follows for M itself, by using the convergence theorem of domains of holomorphy.

In conclusion we will discuss the question of embedding holomorphically complete manifolds in spaces of higher dimension. For this there are several theorems which R. Remmert has recently proved [50]. For example: every n-dimensional holomorphically complete complex manifold M^n can be embedded without singularities in C^N, $N \geq (n+2)^2 - 1$, as a closed analytic submanifold M^n. Thus, the theory of n-dimensional Stein manifolds can also be regarded as the theory of n-dimensional, closed submanifolds of C^N. From this theorem it follows in particular by restricting the Euclidean metric of the C^N on M^n, that on each holomorphically complete complex manifold there exists a complete Kaehler metric with a global potential. This result was already obtained in 1954 using completely different methods by H. Grauert [51].

E. Now, function theory of holomorphically complete manifolds has been extensively developed. The question arises as to how far the function theory on the analytic configurations of functions $f(z_1, \ldots, z_n)$ was included in this development. Oka has recently shown that each unramified domain of existence of a holomorphic function over C^n is holomorphically convex and therefore is a holomorphically complete manifold. This was an exceptional accomplishment which justifies the extensive treatment that holomorphically complete manifolds have received.

F. As mentioned in Section 1 there are, however, ramified analytic configurations of holomorphic functions which have non-uniformizable points. This means that we cannot avoid the discussion of non-uniformizable points. Configurations with such points cannot be complex manifolds. If we want to obtain an equivalence between the domains of holomorphy and our abstract configurations, we have to extend the concept of holomorphically complete manifolds further because of the non-uniformizable ramification points. This leads us again to the concept of complex space. It can be shown that each analytic configuration is a pair $\mathfrak{G} = (G, \Phi)$ with G designating a complex space and Φ a holomorphic, almost one-to-one mapping of G into C^n. The question again arises whether G is holomorphically convex. In contrast to the Oka result for unramified domains of existence, H. Grauert and R. Remmert have shown: there are ramified non-holomorphically convex analytic configurations of holomorphic functions.

However, the investigations of such non-holomorphically convex domains of holomorphy is just beginning. We therefore limit the following considerations of abstract complex spaces and concrete Riemann domains to holomorphically convex configurations. Our goal is to show that with this limitation there exists equivalence between the analytic configurations of holomorphic functions and the abstract holomorphically complete spaces. In order to define the latter concept the axioms of Grauert can easily be carried over to complex spaces. Holomorphic convexity and K-completeness can be defined in the same way for complex spaces as for complex manifolds. There is however not an immediate analogue for each of the Stein axioms. Thus we are lead to the concept of the holomorphically complete space.

Examples of holomorphically complete spaces are naturally the holomorphically convex Riemann domains over C^n since, on account of the axioms, they are K-complete.

Conversely it was shown [52] by construction of an almost one-to-one mapping, that every holomorphically complete space can be holomorphically mapped one-to-one onto a Riemann domain over C^n. We have thus reached the above-mentioned equivalence.

We now prove certain properties of the holomorphically complete space X. There is no difficulty in obtaining a countable base for the open sets because of the equivalence of X with a holomorphically convex analytic configuration over C^n. But what about the separation property of the holomorphic functions in X required as an axiom by Stein? This property can be proved in the same way as for the holomorphically complete manifolds with the help of the Oka linking lemma.

The third Stein property remains: the existence of local coordinates given by functions holomorphic on X. This property is not everywhere meaningful in X (namely not at the non-uniformizable points). A property of equal value can, however, be substituted for it.

For each point $x \in X$ there are finitely many holomorphic functions $f_1, \ldots, f_k$ on X which give a one-to-one mapping of a neighborhood of x onto an analytic set normally embedded in a domain $G \subset C^k$.

Thereby, an analytic set $A \subset G$ is called *normally embedded in a point* $x \in A$, if it is locally irreducible in x and the ring of functions holomorphic on A at x contains only such functions as can be continued holomorphically into a whole k-dimensional neighborhood of x. An analytic set is called *normally embedded* if it is normally embedded in each of its points.

In each holomorphically complete space X the above-mentioned property holds. If X does not possess non-uniformizable points, this property is identical with the existence of the local coordinates required by Stein, which can be given by functions holomorphic in all of X.

Now, function theory can be built up in holomorphically complete spaces. As in the case of the holomorphically complete manifolds, an analogue of the Runge approximation theorem is valid, and analogues to the assertions of Cousin I and Cousin II and the theorem of Poincaré are also valid. Also, the assertions of the Cartan-Serre theory of analytic sheaves can be carried over to holomorphically complete spaces without great difficulty.

5. Plurisubharmonic functions

In 1942, at about the same time but independently, K. Oka and P. Lelong introduced plurisubharmonic functions into the literature [53]. Oka used the new concept to solve, in the case $n = 2$, Levi's well known problem in which the equivalence of pseudoconvex domains and domains of holomorphy was conjectured. For the profound conclusion that in C^2 each pseudoconvex domain $\mathfrak{G}$ is a domain of holomorphy, Oka showed with plurisubharmonic functions that $\mathfrak{G}$ can be exhausted by a sequence of relatively compact pseudoconvex subdomains $\mathfrak{G}_\nu \Subset \mathfrak{G}$ possessing a simpler and smoother boundary than $\mathfrak{G}$. By a carefully constructed linking system he succeeded in proving that all the $\mathfrak{G}_\nu$ are domains of holomorphy. That $\mathfrak{G}$ itself is a domain of holomorphy follows from a theorem of H. Behnke and K. Stein.†

In 1954, H. Bremermann and F. Norguet overcame greater difficulties

† In the complex number space C^n a domain is a domain of holomorphy if it can be exhausted by domains of holomorphy [54].

in extending Oka's proof to the case of more than two complex variables [55]. In the same year Oka was able to show independently that each unramified pseudoconvex domain over C^n is a holomorphically convex domain of holomorphy [56]. This result, too, could not have been achieved without the use of plurisubharmonic functions.

The plurisubharmonic function is a generalization of subharmonic function of one complex variable. A function $s(z)$ on a domain G of the z-plane is said to be subharmonic if it has the following three properties:

(a) Its values are the real numbers and $-\infty$.

(b) It is upper semicontinuous ($\overline{\lim}_{z \to z_0} s(z) \leq s(z_0)$).

(c) If $K \Subset G$ is a closed circle and $h(z)$ a harmonic function on K with $h(z) \geq s(z)$ on the boundary ∂K, it follows that $h(z) \geq s(z)$ also in the interior of K.

The concept of plurisubharmonic function will now be defined by means of subharmonic functions.

A function $p(z_1, \ldots, z_n)$ on a domain $G \subset C^n$ is called *plurisubharmonic* if:

1. $-\infty \leq p(z_1, \ldots, z_n) < +\infty$
2. $p(z_1, \ldots, z_n)$ is upper semi-continuous
3. the restriction of $p(z_1, \ldots, z_n)$ to each part of a one-dimensional analytic plane in G is subharmonic.

P. Lelong was soon able to show that the class of plurisubharmonic functions is invariant with respect to holomorphic coordinate transformations [57]. This property made it meaningful to define the concept of plurisubharmonic functions on complex manifolds. A literal carryover of the definition to the case of complex spaces is however impossible on account of the non-uniformizable points. As soon as it was possible to substitute for Axiom 3 an equivalent, more suitable, and more general requirement 3′, it was possible to meaningfully introduce plurisubharmonic functions also into complex spaces.† The new Axiom 3′ says:

3′. If φ is a holomorphic mapping of the unit circle K of the z-plane into G, the function $q(z) = p \circ \varphi$ is subharmonic in K.

The statement that 3 and 3′ are equivalent includes the property of invariance of plurisubharmonic functions with respect to holomorphic coordinate transformations.

The definition of plurisubharmonic functions on complex spaces can be given in the following manner:

A function $p(x)$ on a complex space X is called plurisubharmonic if:

1. Its values are the real numbers and $-\infty$,
2. It is upper semi-continuous on X,

† For this and the following theorems and definitions see always: Grauert and Remmert [58].

3. If φ is a holomorphic mapping of the unit circle K of the z-plane into X, the function $p \circ \varphi$ is subharmonic in K.

Let N be the set of non-uniformizable points of X. Accordingly, the restriction of each plurisubharmonic function on X to the complex manifold $X - N$ is plurisubharmonic in the sense of Lelong. Since it can be shown that each plurisubharmonic function on $X - N$ can be uniquely plurisubharmonically continued into X, the above definition might be considered to be the only meaningful generalization of the concept of plurisubharmonic functions to complex spaces.

The theorem concerning the continuation of plurisubharmonic functions appears as a special case of a more general theorem derived by Grauert and Remmert, if one notes that according to Cartan-Oka the set of non-uniformizable points of an n-dimensional complex space X is an analytic set of dimension at most $(n - 2)$.† H. Grauert and R. Remmert have shown:

THEOREM 1. Let X be an n-dimensional complex space and A an analytic set of dimension at most $(n - 2)$ in X; then each plurisubharmonic function in $X - A$ can be uniquely continued to a function plurisubharmonic in all of X.

This theorem is analogous to a theorem of B. Riemann which says that no analytic set of dimension $d \leq n - 2$ can be the singularity surface of a holomorphic function. Corresponding to another theorem of B. Riemann it is true that:

THEOREM 2. If A is an analytic set of dimension at most $(n - 1)$ in an n-dimensional complex space X and $p(x)$ a plurisubharmonic function defined on $X - A$, which is bounded from above in a neighborhood of each point $x_0 \in A$, then $p(x)$ can be uniquely continued into A to a function plurisubharmonic on X.

Both theorems have been derived by P. Lelong in a more general form for complex manifolds.

As mentioned at the beginning of this section, Oka has used plurisubharmonic functions in order to show that every unramified pseudo-convex Riemann domain over C^n is a domain of holomorphy. An important link in his chain of reasoning is the following theorem:

An unramified domain $\mathfrak{G}$ over C^n is a domain of holomorphy if and only if there exists a plurisubharmonic function $p(x)$ on $\mathfrak{G}$ which tends towards $+\infty$ as x approaches the ideal boundary of $\mathfrak{G}$.

For complex spaces, as in the case of the concrete Riemann domains over C^n, it might be possible to characterize holomorphically complete spaces by "local" properties similar to pseudo-convexity. For this purpose it will be necessary to carry over a theorem of the above type to complex

† The proof of this statement is contained in the seminar notes of H. Cartan (1953/54).

spaces. It is indeed possible to construct on a holomorphically complete space X a plurisubharmonic function $p(x)$ which tends towards $+\infty$ as x approaches the ideal boundary of X. Furthermore, such a function can be found on every holomorphically convex space.† Conversely, it can be conjectured that a complex space X is a holomorphically convex space if a plurisubharmonic function $p(x)$ exists on X which tends towards $+\infty$ as x approaches the ideal boundary of X.‡

6. Complex-analytic fibre spaces

1. It was Seifert who in 1933 first used fibre spaces extensively for the investigation of the homeomorphism problem for compact 3-dimensional manifolds [59]. Seifert's definition is still very general. Later, it became evident that the concept of the "fibering" is of importance for differential geometry, especially for the investigation of differentiable structures (Stiefel, Hopf). For this purpose a limitation was introduced resulting in the concept of fibre bundles (H. Whitney).

Further, H. Cartan proved in 1950 that fibre bundles can be successfully used in complex function theory [60]. For this purpose it was necessary to introduce analytic notions into the theory of fibre spaces. This led to the definition of the analytic fibre bundle. H. Cartan showed in particular that the classic problems, Cousin I and Cousin II, can be formulated in the terminology of these analytic fibre bundles.

In the meantime special fibre bundles (complex analytic vector bundles) have been applied extensively in function theory on compact and non-compact complex manifolds. Investigations have taken place by Kodaira, Spencer; Cartan, Serre; Atiyah (of algebraic manifolds). Essentially complex line-bundles were used in the formulation and proof of Hirzebruch's generalization of the Riemann-Roch theorem [61].

2. We will give a short definition of fibre bundles, whose base and fibre are complex spaces and whose structure group is a complex Lie group, as follows:§ Let X be a topological space, $\mathfrak{B}$, $\mathfrak{F}$ complex spaces, L a complex Lie group (i.e. complex manifold between whose points a group operation is defined; the group operations are dependent holomorphically on the points of L).

Let L operate holomorphically in $\mathfrak{F}$ [63]. Roughly speaking, this property means the following: The points $l \in L$ are also analytic

† The following fact has to be used: If f is a holomorphic function then $f \cdot \bar{f}$ is a plurisubharmonic function.

‡ Meanwhile one of the authors has proved that the conjecture is true if $p(x)$ is "strongly" plurisubharmonic. If $p(x)$ is only plurisubharmonic, the conjecture is incorrect.

§ For the definition of the general topological fibre bundle see Steenrod [62].

automorphisms of the complex space $\mathfrak{F}$, the group operation $\circ$ in L is the natural combination of these automorphisms, and the mapping $L \times \mathfrak{F} \to \mathfrak{F}$ of the Cartesian product defined by $(l, y) \to l(y)$, $l \in L$, $y \in \mathfrak{F}$, is holomorphic.

If π is a continuous mapping of X onto $\mathfrak{B}$, π generates a "fibering in X": $\mathfrak{F}_r = \pi^{-1}(r)$, $r \in \mathfrak{B}$ is called the *fibre over r*. The fibering induced in X by π is not enough to make X into a fibre bundle. For this, further conditions must be satisfied:

Let $\{W_\iota : \iota \in I\}$ be an open covering of $\mathfrak{B}$ (I an index set), $\psi_\iota : W_\iota \times \mathfrak{F} \to \pi^{-1}(W_\iota)$ a topological and fibre-preserving mapping (this means $\pi \circ \psi_\iota (r, y) \equiv r$, $r \in W_\iota$, $y \in \mathfrak{F}$). ψ_ι gives a one-to-one correspondence between the points $(r, y) \in W_\iota \times \mathfrak{F}$ and the points of $\pi^{-1}(W_\iota)$. Thus the points (r, y) are to be regarded as coordinates in $\pi^{-1}(W_\iota)$. The triple $K_\iota = (W_\iota \times \mathfrak{F}, \psi_\iota, \pi^{-1}(W_\iota))$ is therefore called a *map on X*. (If there is a map K_ι for $\pi^{-1}(W_\iota)$, then all fibres $\mathfrak{F}_r$ in $\pi^{-1}(W_\iota)$ are topologically of space type $\mathfrak{F}$.) If $W_{\iota_1} \cap W_{\iota_2} \neq \emptyset$ (with arbitrary $\iota_1, \iota_2 \in I$), then $\psi_{\iota_1}^{-1} \circ \psi_{\iota_2}$ is a fibre-preserving mapping of $(W_{\iota_1} \cap W_{\iota_2}) \times \mathfrak{F}$ onto itself. K_{ι_1} and K_{ι_2} are called *compatible* if $\psi_{\iota_1}^{-1} \circ \psi_{\iota_2}$ can be given in the form $r^* = r$, $y^* = \Phi_{\iota_1\iota_2}(r) \,\square\, y$, $r \in W_{\iota_1\iota_2}$, $y \in \mathfrak{F}$. Here $\Phi_{\iota_1\iota_2}(r)$ is a continuous mapping of $W_{\iota_1} \cap W_{\iota_2}$ into L, and $\square$ is the application of the automorphism $\Phi_{\iota_1\iota_2}(r)$ on $y \in \mathfrak{F}$.

If all K_ι are pairwise compatible, then X becomes a fibre bundle through π, $\mathfrak{B}$, $\mathfrak{F}$, L, $\{K_\iota\}$. $\mathfrak{B}$ is called the *base* of X, $\mathfrak{F}$ the *fibre*, L the *structure group*, π the *projection*, and $\{K_\iota\}$ the *bundle structure*.

If all $\Phi_{\iota_1\iota_2}(r)$ are holomorphic, the maps K_ι are called *holomorphically compatible*. The complex structure transformed by ψ_{ι_1} from $W_{\iota_1} \times \mathfrak{F}$ to $\pi^{-1}(W_{\iota_1})$ is now identical in $\pi^{-1}(W_{\iota_1} \cap W_{\iota_2})$ with the similarly obtained complex structure in $\pi^{-1}(W_{\iota_2})$: X carries a natural complex structure and is therefore a complex space. In this case X is called an *analytic fibre bundle* over $\mathfrak{B}$.

Fibre bundles which are structurally equal are called equivalent. Let X_1 and X_2 be two fibre bundles over $\mathfrak{B}$ with the same fibres and the same structure groups.

DEFINITION: X_1 and X_2 are *equivalent* if and only if there exists a fibre-preserving topological mapping $\psi : X_1 \xrightarrow{\text{onto}} X_2$ which can be given in the local fibre-coordinate systems in the form $r^* = r$, $y^* = \Phi_\iota(r) \,\square\, y$. Here $\Phi_\iota(r)$ is a continuous mapping into L.

X_1 and X_2 are called *analytically equivalent* if ψ can be chosen as a holomorphic mapping $X_1 \to X_2$ ($\Phi_\iota(r)$ holomorphic!). This property is meaningful only if X_1 and X_2 carry complex structures, i.e. they are analytic fibre bundles.

3. The simplest example of an analytic fibre space is the Cartesian product $\mathfrak{B} \times \mathfrak{F}$; fibre bundles equivalent to $\mathfrak{B} \times \mathfrak{F}$ are called trivial

(analytically or only continuously trivial!). If $\mathfrak{F}$ is C^q and L the group $Gl(q, C)$ of complex homogeneous linear mappings $\mathfrak{z} \to A \cdot \mathfrak{z}$ of C^q, then X is called a *vector bundle*. In this case the fibres over $r \in \mathfrak{B}$ carry the vector space structure of C^q because of the linear structure group. In the case of $q = 1$, X is called a *complex line bundle*.

As H. Cartan has shown,† each Cousin-II-distribution in $\mathfrak{B}$ is canonically corresponded to a complex-analytic line bundle X over $\mathfrak{B}$. X is continuously trivial if and only if there exists a continuous solution of the Cousin-II-distribution. There exists a holomorphic solution if and only if X is analytically trivial. Oka proved that if $\mathfrak{B}$ is a domain of holomorphy in C^n, then each Cousin-II-distribution in $\mathfrak{B}$ is holomorphically solvable, if and only if there exists a continuous solution. From this it follows that in the case where $\mathfrak{B}$ is a domain of holomorphy, if X is topologically trivial, then X is also analytically trivial.

This result of Oka has been generalized for arbitrary complex-analytic fibre bundles during the past year [68]:

THEOREM Ia. Let $\mathfrak{B}$ be a holomorphically complete space and X a continuously trivial, complex-analytic fibre bundle over $\mathfrak{B}$. Then X is also analytically trivial.

Going much farther than the statement of this theorem, one can even establish an isomorphism between the (continuous) equivalence classes of topological fibre bundles and the (analytic) equivalence classes of analytic fibre bundles:

THEOREM I. Let $\mathfrak{B}$ be a holomorphically complete space, X_1 and X_2 two analytic fibre bundles over $\mathfrak{B}$ with the same structure group and fibre. Then X_1 and X_2 are analytically equivalent if and only if they are continuously equivalent.

THEOREM II. Let $\mathfrak{B}$ be a holomorphically complete space and X a continuous fibre bundle over $\mathfrak{B}$. Then there exists an analytic fibre bundle X^* over $\mathfrak{B}$ with the same structure group and fibre which is equivalent to X.‡

It follows from the isomorphisms obtained in I and II that various statements about topological fibre spaces are also valid in the theory of analytic fibre bundles (for instance, theorems about the restriction of the structure group L of X to subgroups of L). Since one can make use of analytic fibre bundles for complex analysis, it follows that from topological theorems one can derive theorems about complex function theory in holomorphically complete spaces. Such applications have been made in various ways [69].

† See for this and the following theorems: Grauert [66].

‡ The Theorems I and II for complex-analytic fibre bundles with solvable structure group were first proved by J. Frenkel [67].

References

[1] Hartogs, Fritz, Einige Folgerungen aus der Cauchyschen Integralformel bei Funktionen mehrerer Veränderlichen, *Sitzb. Münchener Akad.* **36** (1906), p. 223.

[2] Thullen, Peter, Zur Theorie der Singularitäten der Funktionen zweier komplexer Veränderlichen. —Die Regularitätshüllen, *Math. Annalen* **106** (1932), pp. 64–76.

[3] Behnke, Heinrich und Thullen, Peter, Theorie der Funktionen mehrerer komplexer Veränderlichen, *Erg. d. Math. III*, 3, reprinted by Chelsea Publishing Co., N. Y.

[4] Oka, Kiyoshi, Sur les fonctions analytiques de plusieurs variables. IX—Domaines finis sans point critique intérieur. *Japanese Journal of Mathematics* **23** (1953), pp. 97–155.

[5] *Colloque sur les fonctions de plusieurs variables.* Bruxelles (11.–14. mars 1953), 157 pp. Liège et Paris.

[6] Behnke, Heinrich, Funktionentheorie auf komplexen Mannigfaltigkeiten, *Proceedings of the International Congress of Mathematicians 1954 (Amsterdam)*, **3**, pp. 45–57.

[7] Behnke, Heinrich und Stein, Karl, Modifikation komplexer Mannigfaltigkeiten und Riemannsche Gebiete, *Math. Annalen* **124** (1951), pp. 1–16.

[8] Cartan, Henri, *Séminaire E.N.S.*, 1951–1952.
Séminaire E.N.S., 1953–1954. Paris (multigraphed).

[9] cf. [7].

[10] Grauert, Hans und Remmert, Reinhold, Singularitäten komplexer Mannigfaltigkeiten und Riemannsche Gebiete, *Math. Zeitschrift* **67** (1957), pp. 103–128.

[11] Heinz, Erhard, Ein elementarer Beweis des Satzes von Radó-Behnke-Stein-Cartan über analytische Funktionen, *Math. Annalen* **131** (1956), pp. 258–259.

[12] Remmert, Reinhold und Stein, Karl, Über die wesentlichen Singularitäten analytischer Mengen, *Math Annalen* **126** (1953), pp. 263–306.

[13] cf. [10].

[14] cf. [4].

[15] Remmert, Reinhold, Projektionen analytischer Mengen, *Math. Annalen* **130** (1956), pp. 410–441.
Holomorphe und meromorphe Abbildungen komplexer Räume, *Math. Annalen* **133** (1957), pp. 328–370.

[16] Serre, Jean Pierre, Géométrie algébrique et géométrie analytique, *Annales de l'Institut Fourier* **VI** (1955 et 1956).

[17] Cartan, Henri, Détermination des points exceptionels d'un système de p fonctions analytiques de n variables complexes, *Bull. Sci. Math.* **57** (1933), pp. 334–344;
and Remmert, Reinhold, Holomorphe und meromorphe Abbildungen komplexer Räume, *Math. Annalen* **133** (1957), pp. 328–370.

[18] REMMERT, REINHOLD, Meromorphe Funktionen in kompakten komplexen Räumen, *Math. Annalen* **132** (1956), pp. 277–288.

[19] REMMERT, REINHOLD, Holomorphe und meromorphe Abbildungen komplexer Räume, *Math. Annalen* **133** (1957), pp. 328–370.

[20] cf. [19].

[21] cf. [19].

[22] STEIN, KARL, Analytische Zerlegungen komplexer Räume, *Math. Annalen* **132** (1956), pp. 68–93.

[23] cf. [22].

[24] cf. [22].

[25] cf. [22].

[26] REMMERT, REINHOLD, Reduktion komplexer Räume, To appear in *Math. Annalen.*

[27] KOCH, KARL, Die analytische Projektion, *Schriftenreihe des Math. Instituts d. Univ. Münster*, Heft 6 (1953).

[28] STEIN, KARL, *Analytische Projektion komplexer Mannigfaltigkeiten.* Colloque Bruxelles (cf. [5]), pp. 87–107.

[29] cf. [7].

[30] ZARISKI, OSCAR, Reduction of the Singularities of Algebraic Three-Dimensional Varieties, *Annals of Mathematics* **45** (1944), pp. 472–542.

[31] HIRZEBRUCH, FRIEDRICH, Über vierdimensionale Riemannsche Flächen mehrdeutiger analytischer Funktionen von zwei komplexen Veränderlichen, *Math. Annalen* **126** (1953), pp. 1–22.

[32] KREYSZIG, ERWIN, Stetige Modifikationen komplexer Mannigfaltigkeiten, *Math. Annalen* **128** (1955), pp. 479–492.

REMMERT, REINHOLD, Über stetige und eigentliche Modifikationen komplexer Räume, *Colloque de topologie de Strasbourg* (Dec. 1954), multigraphed.

GRAUERT, HANS und REMMERT, REINHOLD, Zur Theorie der Modifikationen. I. Stetige und eigenliche Modifikationen komplexer Räume, *Math. Annalen* **129** (1955), pp. 274–296.

STOLL, WILHELM, Über meromorphe Modifikationen.

I. Allgemeine Eigenschaften der Modifikationen, *Math. Zeitschrift* **61** (1954), pp. 206–234.

II. Allgemeine Eigenschaften der meromorphen Modifikationen, *Math. Zeitschrift* **61** (1955), pp. 467–488.

III. Streueigenschaften analytischer und meromorpher Modifikationen, *Math. Zeitschrift* **62** (1955), pp. 189–210.

IV. Die Erzeugung analytischer und meromorpher Modifikationen zwischen kompakten Mannigfaltigkeiten durch σ-Prozesse, *Math. Annalen* **130** (1955), pp. 147–182.

V. Die Erzeugung analytischer und meromorpher Modifikationen durch σ-Prozesse, *Math. Annalen* **130** (1955), pp. 272–316.

[33] cf. §4 p. 32.

[34] cf. §4 p. 32.

[35] cf. [6].

[36] BEHNKE, HEINRICH und STEIN, KARL, Entwicklung analytischer Funktionen auf Riemannschen Flächen, *Math. Annalen* **120** (1947), pp. 430–461.

[37] CALABI, EUGENIO and ECKMANN, BENO, A class of compact complex manifolds, which are not algebraic, *Ann. of Math.* **58** (1953), pp. 494–500.

[38] STEIN, KARL, Analytische Funktionen mehrerer komplexer Veränderlichen und das zweite Cousinsche Problem, *Math. Annalen* **123** (1951), pp. 201–222.

[39] BEHNKE, HEINRICH, Généralisation du théorème de RUNGE pour les fonctions multiformes de variables complex. cf. [5].

[40] CARTAN, HENRI, Variétés analytiques complexes et cohomologie.
SERRE, JEAN PIERRE, Quelques problèmes globaux relatifs aux variétés de STEIN.
Both cf. [5].

[41] GRONWALL, T. H., On the expressibility of a uniform function of several complex variables as the quotient of two functions of entire character, *American M.S. Trans.* **18** (1917), pp. 50–64.

[42] OKA, KIYOSHI, Sur les fonctions de plusieurs variables. II. Domaines d'holomorphie, *Journ. of Sci. Hiroshima Univ. Ser. A*, Vol. 7, Nr. 2 (1937).

[43] STEIN, KARL, Topologische Bedingungen für die Existenz analytischer Funktionen komplexer Veränderlichen zu vorgegebenen Nullstellen, *Math. Annalen* **117** (1940), pp. 727–757.

[44] HIRZEBRUCH, FRIEDRICH, Neue topologische Methoden in der algebraischen Geometrie, *Erg. d. Math., Neue Folge*, Heft 9, (1956).

[45] cf. [40].

[46] cf. [5].

[47] RADÓ, TIBOR, Über den Begriff der Riemannschen Fläche, *Acta Szeged* **2** (1924), pp. 101–121.

[48] CALABI, EUGENIO and ROSENLICHT, MAXWELL, Complex analytic manifolds without countable base, *Proc. of the Amer. Math. Soc.* **4** (1953), pp. 335–340;
and HOPF, HEINZ, Schlichte Abbildungen und lokale Modifikationen 4-dimensionaler komplexer Mannigfaltigkeiten, *Comment. Math. Helv.* **29** (1955), pp. 132–156.

[49] GRAUERT, HANS, Charakterisierung der holomorph vollständigen Räume, *Math. Annalen* **129** (1955), pp. 233–255.

[50] REMMERT, REINHOLD, Will be published in *Math. Annalen.*

[51] GRAUERT, HANS, Charakterisierung der Holomorphiegebiete durch die vollständige Kählersche Metrik, *Math. Annalen* **131** (1956), pp. 38–75.

[52] cf. [49].

[53] LELONG, PIERRE, *C. R. Acad. Sci. Paris* **125** (1942), pp. 398 and 454.
and OKA, KIYOSHI, Sur les fonctions analytiques de plusieurs variables. VI. Domaines pseudo-convexes, *Tohoku Math. Journal* **49** (1942), pp. 15–52.

[54] BEHNKE, HEINRICH AND STEIN, KARL, Konvergente Folgen nichtschlichter Regularitätsbereiche, *Ann. Mat. pura appl.* **28** (1949).

[55] BREMMERMANN, HANS-JOACHIM, Über die Äquivalenz der pseudokonvexen Gebiete und der Holomorphiegebiete im Raum von n komplexen Veränderlichen, *Math. Ann.* **128** (1954), pp. 63–91;

and NORGUET, FRANÇOIS, Sur les domaines d'holomorphie des fonctions uniformes de plusieurs variables complexes, *Bull. Soc. Math. France* **82** (1954), pp. 139–159.

[56] cf. [4].

[57] LELONG, PIERRE, Fonctions plurisousharmoniques; mesures de Radon associées. Applications aux fonctions analytiques, cf. [5].

[58] GRAUERT, HANS und REMMERT, REINHOLD, Plurisubharmonische Funktionen in komplexen Räumen, *Math. Zeitschr.* **65** (1956), pp. 175–194.

[59] SEIFERT, H., Topologie dreidimensionaler gefaserter Räume, *Acta Math.* **60** (1933), pp. 147–238.

[60] CARTAN, HENRI, Espaces fibrés analytiques complexes, *Séminaire Bourbaki* (1950).

[61] cf. [44].

[62] STEENROD, NORMAN, *The Topology of Fibre Bundles*, Princeton, 1951.

[63] For exact definition see: F. Hirzebruch, *loc. cit.* [44].

[64] cf. CARTAN [59].

[65] OKA, KIYOSHI, Sur les fonctions analytiques de plusieurs variables. III. Deuxième problème de Cousin, *Journ. of Sci. Hirosima Univ., Ser. A*, **19** (1938), pp. 7–19.

[66] GRAUERT, HANS, Analytische Faserungen über holomorph vollständigen Räumen. To appear in *Math. Annalen* 1958.

[67] FRENKEL, JEAN, Sur une classe d'espaces fibrés analytiques, *C. R. Acad. Sci. Paris* **236** (1953), pp. 40–41.
Sur les espaces fibrés analytiques complexes de fibre résoluble, *C. R. Acad. Sci. Paris* **241** (1955), pp. 16–18.

[69] See for instance:
RÖHRL, HELMUT, Das Riemann-Hilbertsche Problem der Theorie der linearen Differentialgleichungen, *Math. Annalen* **133** (1957), pp. 1–25;
GRAUERT, HANS, Das Levische Problem für unverzweigte Riemannsche Gebiete über Steinschen Mannigfaltigkeiten. (Lecture: Oberwolfach, March 1957);
DOCQUIER, FERDINAND, Der Rungesche Satz in Steinschen Mannigfaltigkeiten, (Dissertation Münster 1957).

The Complex Analytic Structure of the Space of Closed Riemann Surfaces

Lars V. Ahlfors†
HARVARD UNIVERSITY

1. Introduction

1.1. In the classical theory of algebraic curves many attempts were made to determine the "modules" of an algebraic curve. The problem was vaguely formulated, and the only tangible result was that the classes of birationally equivalent algebraic curves of genus $g > 1$ depend on $6g-6$ real parameters. More recent attempts to go to the bottom of the problem by more powerful algebraic methods have also ended in failure.

The corresponding transcendental problem is to study the space of closed Riemann surfaces and, if possible, introduce a complex analytic structure on that space. In this direction considerable progress has been made. The most important step was taken by Teichmüller [3] who realized that the problem becomes much simpler if the notion of conformal equivalence is replaced by a weaker equivalence relation in which the conformal mappings are subject to topological restraints. Roughly speaking, this leads to a covering space whose intrinsic properties are easier to handle. For the original problem it would still be necessary to study the ramifications of the covering space. The latter problem is of a completely different nature, and Teichmüller's approach thus has the advantage of allowing us to attack one difficulty at a time.

In his famous paper of 1940 Teichmüller formulated a bold program based on the theory of extremal quasiconformal mappings [3]. The paper was strongly heuristic, but in 1943 he managed to prove the main conjecture, albeit in a heavyhanded manner [4]. The proof has since been simplified [1], but remains less direct than one would desire. In any case, the theorem is firmly established, and it shows that the covering space has a natural metric in which it is homeomorphic to Euclidean space of

† A considerable part of the research was carried out under Contract AF 18(600)-1461.

dimension $6g-6$. The topology is the same that one would obtain by considering variable branch-points or the parameters in the generators of the Fuchsian group.

1.2. The classical problem calls for a complex analytic structure rather than a metric. Teichmüller states explicitly, in 1944, that his metrization is of no use for the construction of an analytic structure [5]. The present author disagrees and will show that the metrization and the corresponding parametrization are at least very convenient tools for setting up the desired structure.

The problem is not a clear cut one, and several formulations seem equally reasonable. In his paper of 1944, one of his last, Teichmüller analyzes the situation and ends by setting his sights extremely high. He does not claim complete success, and due to the sketchy nature of the paper I have been unable to determine just how much he has proved.

My own goal is more modest, for I am willing to settle for any intrinsic formulation that leads to existence and uniqueness. To be specific, my starting point is the requirement that the periods of the normalized Abelian integrals of the first kind should be analytic functions. Subject to a mild topological restriction, we intend to prove that there is one and only one complex analytic structure which satisfies this requirement.

2. Formulation of the problem

2.1. We have already remarked that we shall use an equivalence relation which is weaker than conformal equivalence. In order to define this equivalence we need to consider the fundamental groups.

Let X be a Riemann surface of genus g. We choose a point $p \in X$ and denote by $\pi_1(X, p)$ the fundamental group formed by the homotopy classes of closed curves from p. If q is another point on X, we know that $\pi_1(X, q)$ is isomorphic to $\pi_1(X, p)$. What is more, we can associate a unique isomorphism with every arc σ from p to q: it is the isomorphism T^σ which transforms the homotopy class of a closed curve c from p into the homotopy class of $\sigma^{-1}c\sigma$. If $p = q$, then T^σ is an inner automorphism, and conversely every inner automorphism corresponds to a closed curve from p.

The group $\pi_1(X, p)$ can be generated by $2g$ generators $a_1, b_1, \ldots, a_g, b_g$ which satisfy the single relation $a_1b_1\ a_1^{-1}b_1^{-1} \ldots a_gb_ga_g^{-1}b_g^{-1} = 1$. Any such ordered system of generators is called a *canonical system*.

Let Σ_p and Σ_q be canonical systems for $\pi_1(X,p)$ and $\pi_1(X,q)$ respectively. We shall say that Σ_q is equivalent to Σ_p if and only if $\Sigma_q = T^\sigma\ \Sigma_p$ for some σ; the notation means that each generator in Σ_p is transformed by T^σ into the corresponding generator in Σ_q. The conditions for an equivalence relation are obviously satisfied.

Suppose now that ϕ is a topological mapping of X onto another surface X'. Then any canonical system Σ on X is transformed into a canonical system $\phi(\Sigma)$ on X' formed by the images of the generators. We shall say that ϕ maps (X, Σ) onto (X', Σ') if and only if $\phi(\Sigma)$ is equivalent to Σ'. Finally, if ϕ is a directly conformal mapping we say that ϕ maps (X, Σ) conformally onto (X', Σ'), and the two pairs are said to be conformally equivalent. Again, the conditions for an equivalence relation are trivially fulfilled.

Observe that by this definition (X, Σ) and (X, Σ') are conformally equivalent under the identity mapping if and only if Σ and Σ' are equivalent generator systems.

A class $\mathfrak{X}$ of conformally equivalent pairs (X, Σ) would be most correctly referred to as an *abstract Riemann surface with generators*. We shall use the shorter name *canonical Riemann surface*. Any individual pair $(X, \Sigma) \in \mathfrak{X}$ is a realization of the canonical surface $\mathfrak{X}$. For given g the space of all canonical Riemann surfaces will be denoted by M_g.

2.2. Given X and $\Sigma = \{a_1, b_1, \ldots, a_g, b_g\}$ there exists a unique analytic differential $\theta_i dz$ with period 1 along a_i and periods 0 along all a_k, $k \neq i$. The period of θ_i along b_j is denoted by τ_{ij}. These numbers are elements of the Riemann matrix $\tau(X, \Sigma)$ associated with X and Σ.

When X and Σ are transformed by the same conformal mapping it is clear that the matrix τ remains unchanged. The same is true if Σ is replaced by an equivalent system Σ'. Hence $\tau(X, \Sigma)$ depends only on the equivalence class of (X, Σ). In other words, the elements τ_{ij} are well defined complex-valued functions on the space M_g.

2.3. The definition of a complex analytic structure is very familiar and need not be repeated. We recall merely that it is customary to superimpose the analytic structure on a topological structure which is already present. If the topological space is connected and able to carry a complex analytic structure, it is necessarily a manifold of even dimension equal to twice the complex dimension.

We have already mentioned that Teichmüller has discovered a natural metric which topologizes M_g as a $(6g-6)$-dimensional manifold. We shall use this topology as the underlying topology. Our aim is thus to prove:

THEOREM. *The space M_g of canonical Riemann surfaces of genus $g > 1$ permits one and only one complex analytic structure over the Teichmüller topology with respect to which all elements τ_{ij} of the Riemann matrix are analytic functions. The complex dimension of this structure is $3g-3$.*

I have not succeeded in showing that the Teichmüller topology is the only topology which permits a complex analytic structure of the desired kind. Nevertheless, there are sufficient indications that the Teichmüller topology is the only one that it is reasonable to consider.

3. Teichmüller's theorem

3.1. We shall make extensive use of a result that was first formulated and proved by Teichmüller in connection with his theory of extremal quasiconformal mappings [3, 4]. This result leads to a parametrization of M_g as a Euclidean space, and also to a metric structure which is compatible with the Euclidean topology.

We state Teichmüller's theorem in two parts. The first part is an existence and uniqueness theorem that does not refer explicitly to the quasiconformal properties.

Let $\tilde{X}_0$ and $\tilde{X}$ be distinct points of M_g, and choose $(X_0, \Sigma_0) \in \tilde{X}_0$, $(X, \Sigma) \in \tilde{X}$. With this choice there are associated: (1) *a positive number $k < 1$,* (2) *a pair of analytic quadratic differentials fdz^2 and $\phi d\zeta^2$ on X_0 and X respectively,* (3) *a topological mapping of (X_0, Σ_0) onto (X, Σ). If z and ζ are local coordinates of points that correspond under the mapping, then*

$$\sqrt{\phi}d\zeta = \sqrt{f}dz + k\sqrt{\bar{f}}d\bar{z}. \tag{1}$$

The mapping and the constant k are uniquely determined, while f and ϕ are determined up to a common positive factor.

There are several points in this statement that require further clarification. In the first place, in (1) it is understood that $\sqrt{\bar{f}}$ denotes the complex conjugate of $\sqrt{f}$, and the sign of $\sqrt{f}$ will determine the sign of $\sqrt{\phi}$. Without ambiguity (1) could be replaced by the equivalent relation

$$\phi d\zeta^2 = fdz^2 + 2k|f|dzd\bar{z} + k^2\bar{f}d\bar{z}^2, \tag{1'}$$

but we prefer (1) because of its more manageable form.

Secondly, the zeros of f will be mapped upon zeros of ϕ with the same multiplicity. At these points the mapping is actually not differentiable, so that (1) and (1′) can be regarded as fulfilled only by convention. In a neighborhood of any other point the mapping can be expressed by a function $\zeta = \zeta(z)$ whose complex derivatives ζ_z, $\zeta_{\bar{z}}$ exist and satisfy

$$\begin{aligned} \phi(\zeta(z))\zeta_z^2 &= f(z) \\ \phi(\zeta(z))\zeta_{\bar{z}}^2 &= k^2 f(z). \end{aligned} \tag{1''}$$

It should be noted that these equations are invariant under changes of both local variables.

3.2. It is clear that a conformal mapping changes all quadratic differentials according to a canonical correspondence. If we replace (X_0, Σ_0) and (X, Σ) by other representatives of the same equivalence classes, it follows from the uniqueness part of Teichmüller's theorem that fdz^2 and $\phi d\zeta^2$ will change canonically, while k remains the same, and the mapping is merely composed with the conformal mappings. In other words, Teichmüller's mapping may be thought of as associated with $\tilde{X}_0$ and $\tilde{X}$.

Conversely, let the base surface (X_0, Σ_0) remain fixed and change (X, Σ) to (X_1, Σ_1). We use a subscript to indicate the change and assume that $k_1 = k$ and $f_1 = f$ (after multiplication with a positive constant). Then the induced mapping $\zeta \rightarrow \zeta_1$ satisfies $\phi d\zeta^2 = \phi_1 d\zeta_1^2$; hence it is conformal except perhaps at the zeros of ϕ, and because of the continuity it will be conformal everywhere. We conclude that (X, Σ) is conformally equivalent to (X_1, Σ_1), and hence that $\tilde{X}$ is uniquely determined by f and k. Incidentally, Teichmüller's theorem implies that the only conformal mapping of (X, Σ) on itself is the identity.

We show further that f and k are completely arbitrary. To see this we introduce a new complex structure on X_0 in which a function is said to be analytic if it can be expressed as an analytic function of the variable

$$\zeta^* = \int_{z_0}^{z} \sqrt{f}dz + k\int_{z_0}^{z} \sqrt{\bar{f}}d\bar{z}.$$

At a point where $f(z_0) \neq O$ the variable ζ^* will indeed yield a topological local mapping. If z_0 happens to be an m-fold zero of f the definition of ζ^* must be modified to

$$\zeta^* = \left(\int_{z_0}^{z} \sqrt{f}dz + k\int_{z_0}^{z} \sqrt{\bar{f}}d\bar{z}\right)^{2/m+2}.$$

With this structure X_0 becomes a Riemann surface that we denote by $X_0(f, k)$. The identity mapping is the Teichmüller mapping that connects (X_0, Σ_0) and $(X_0(f, k), \Sigma_0)$. In fact, $d\zeta^{*2} = (\sqrt{f}dz + k\sqrt{\bar{f}}d\bar{z})^2$ will be an analytic quadratic differential on $X_0(f, k)$.

3.3. We have found a one-one correspondence between all points $\tilde{X} \neq \tilde{X}_0$ on M_g and the pairs (k, f) formed by a positive number $k < 1$ and a quadratic differential fdz^2 on X_0, except that differentials which differ by a positive constant factor correspond to the same $\tilde{X}$.

It is well known that the quadratic differentials form a linear space of dimension $N = 3g-3$ over the complex numbers. Thus we can find N linearly independent $f_1, \ldots, f_N$, and any quadratic differential fdz^2 can be uniquely expressed through $f = c_1f_1 + \ldots + c_Nf_N$. Since a positive factor is irrelevant we can normalize by requiring that $|c_1|^2 + \ldots + |c_N|^2 = 1$.

We introduce the N complex numbers $u_1 = kc_1, \ldots, u_N = kc_N$ as parameters of (X, Σ). They are subject to the condition $0 < |u_+|^2 + \ldots + |u_N|^2 < 1$, and the value of k is obtained from $k^2 = |u_1|^2 + \ldots + |u_N|^2$. The values $(0, \ldots, 0)$ can be assigned to the base surface (X_0, Σ_0). In this way we have established a one-one correspondence between the abstract surfaces with generators and the points of a solid sphere in $2N$ real dimensions.

We remark that the parametrization depends very essentially on the choice of the base point $\tilde{X}_0$. The transition from one parameter system to another is by no means simple, and the study of these parameter mappings will indeed be our main concern.

3.4. The second part of Teichmüller's theorem can be stated as follows: *The Teichmüller mapping of* (X_0, Σ_0) *on* (X, Σ) *is quasiconformal with constant dilatation* $K = (1+k)/(1-k)$. *Among all quasiconformal mappings it is the one whose maximal dilatation is the least.*

We define the *distance* between $\tilde{X}_0$ and $\tilde{X}$ to be $d(\tilde{X}_0, \tilde{X}) = \log K$. Since reciprocal mappings have the same maximal dilatation it is clear that $d(\tilde{X}, \tilde{X}_0) = d(\tilde{X}_0, \tilde{X})$. Furthermore, if two mappings are composed, the dilatation of the composed mapping is at most equal to the product of the dilatations. This proves the triangle inequality. Finally, the maximal dilatation is 1 if and only if the mapping is conformal. It follows that the distance $d(X_0, X)$ defines M_g as a metric space. We shall refer to this metric space as the *Teichmüller space*.

It is known that the parameters $(u_1, \ldots, u_N)$ are continuous in the Teichmüller metric, and indeed that the parameter mapping is a homeomorphism. Hence the transitions from one parameter system to another are likewise continuous.

4. Hyperelliptic surfaces

4.1. We shall find that the distinction between hyperelliptic and non-hyperelliptic surfaces is essential for the line of reasoning that we are going to follow. The need for this distinction has its root in the following classical theorem of M. Noether: If X is a non-hyperelliptic surface, every quadratic differential $f dz^2$ on X can be expressed in the form

$$f = \Sigma\, c_{ij} \theta_i \theta_j, \tag{2}$$

where the c_{ij} are constants, and each $\theta_i dz$ is an Abelian differential of the first kind.

In contrast, if X is hyperelliptic, the maximum number of linearly independent products $\theta_i \theta_j$ is only $2g-1$. The numbers $3g-3$ and $2g-1$ coincide for $g = 2$, corresponding to the fact that all surfaces of genus 2 are hyperelliptic. In that case the representation (2) is of course always possible.

4.2. A hyperelliptic surface of genus g can be represented as the two-sheeted Riemann surface of $\sqrt{Q(z)}$ where

$$Q(z) = (z - z_1) \ldots (z - z_{2g+2})$$

is a polynomial of degree $2g+2$. The differentials of the first kind are given by

$$\frac{P(z)dz}{\sqrt{Q(z)}}, \tag{3}$$

where P is a polynomial of degree $\leqq g-1$. Their products generate $2g-1$ linearly independent quadratic differentials of the form

$$\frac{A(z)dz^2}{Q(z)},$$

$\deg A \leqq 2g-2$. The remaining quadratic differentials which cannot be so represented are of the form

$$\frac{B(z)dz^2}{\sqrt{Q(z)}},$$

with $\deg B \leqq g-3$.

More precisely, every quadratic differential is a sum of an even and an odd quadratic differential. The even ones are generated by the products $\theta_i\theta_j$, but the odd ones cannot be generated in this way.

4.3. There is essentially only one way in which a hyperelliptic surface can be represented as a two-sheeted covering surface of the sphere. This is very familiar, but for the sake of completeness we include a proof. Let the interchange of sheets in one such representation be denoted by J. If F is a function on the surface, then $(F(p) - F(Jp))^2$ is a rational function of z. It has at least $2g+2 > 4$ zeros, namely one zero at each branch-point. Unless it is identically zero it must have equally many poles, and this is impossible if F has only two poles. Therefore any function of order two satisfies $F(p) = F(Jp)$, and we conclude that the involution defined by F coincides with J. Moreover, F is a linear function of z.

Let us also prove that any surface of genus g which permits a conformal involution J with $2g+2$ fixpoints is hyperelliptic. If we identify p and Jp, the surface appears as a two-sheeted covering surface with $2g+2$ branch-points of a closed surface whose genus we denote by g_0. The Riemann-Hurwitz relation implies $2g-2 = 2(2g_0 - 2) + 2g + 2$. Hence $g_0 = 0$, so that the surface is hyperelliptic and J corresponds to the interchange of sheets.

4.4 Let X_0 and X be hyperelliptic surfaces with involutions J_0 and J. We shall say that the pairs (X_0, Σ_0) and (X, Σ) are hyperelliptically equivalent if the mappings of X_0 on X which take Σ_0 into Σ will also transform $J_0\Sigma_0$ into a system equivalent to $J\Sigma$. We show that this equivalence relation is compatible with conformal equivalence. For this purpose, suppose that S is a conformal mapping of (X_0, Σ_0) on (X, Σ). Then $S^{-1}JS$ is a conformal involution on X_0 with $2g+2$ fixpoints. By what we have just seen, this implies $S^{-1}JS = J_0$. This is precisely the condition for hyperelliptic equivalence.

We denote by H_g the set of all $\tilde{X}$ such that X is hyperelliptic. We have just shown that H_g permits a decomposition into hyperelliptic equivalence classes.

4.5. Let $f dz^2$ be an even quadratic differential on X_0. We consider the Teichmüller mapping T which corresponds to this f and an arbitrary k; the image surface is called X. Because $f(J_0 z) = f(z)$ and $dJ_0 z = dz$, it is readily seen that the mapping TJ_0T^{-1} is conformal on X. It is an involution with $2g+2$ fixpoints. Therefore X is hyperelliptic with the involution $J = TJ_0T^{-1}$. Thus (X_0, Σ_0) and $(X, T\Sigma_0)$ are hyperelliptically equivalent.

Conversely, assume that (X, Σ) is hyperelliptically equivalent to (X_0, Σ_0), and consider the Teichmüller mapping. Because of the uniqueness of the Teichmüller mapping we know that $J_0 z$ must correspond to $J\zeta$. For this reason we have simultaneously

$$\sqrt{\phi(\zeta)}d\zeta = \sqrt{f(z)}dz + k\sqrt{\bar{f}(\bar{z})}d\bar{z},$$

and
$$\sqrt{\phi(J\zeta)}dJ\zeta = \sqrt{f(J_0 z)}dz + k\sqrt{\bar{f}(\overline{J_0 z})}d\bar{z}.$$

From these relations it is seen that the ratio of the right-hand members is independent of the direction of dz, and that is possible only so that $f(J_0 z)/f(z)$ is positive, and hence equal to a constant $C > 0$. The involutory nature of J_0 yields $C^2 = 1$. Hence $C = 1$, and we have proved that $f dz^2$ is even.

In other words, the surfaces corresponding to an even f on the base surface (X_0, Σ_0) are exactly those which are hyperelliptically equivalent to the base surface.

4.6. It is trivial from what we have seen that each hyperelliptic equivalence class is closed and connected in the Teichmüller topology. We intend to show that the equivalence classes are the components of H_g. Moreover, H_g is closed and the components are isolated.

Let (X_0, Σ_0) represent a point $\tilde{X}_0 \in H_g$. If X is hyperelliptic and (X, Σ) is sufficiently close to (X_0, Σ_0) in the Teichmüller metric, we contend that (X, Σ) is hyperelliptically equivalent to (X_0, Σ_0). Let the Teichmüller mapping of (X_0, Σ_0) onto (X, Σ) be T, and its dilatation K. The mapping $T^{-1}JT$ has dilatation at most K^2. But it is known that the points $\tilde{X}_0$ which correspond to the same surface X_0 are isolated in the Teichmüller metric (for a proof, see Jenkins, [2] p. 183). Therefore, as soon as k is sufficiently small, $T^{-1}JT$ is conformal. Since it is an involution, we conclude as before that $T^{-1}JT = J_0$, and our contention is proved. We have shown that the equivalence classes are relatively open in H_g, and thus that they are isolated components.

The same reasoning shows that H_g is closed. Without supposing X_0

to be hyperelliptic we assume that there are hyperelliptic X such that (X, Σ) is arbitrarily close to (X_0, Σ_0). With the same notation as before $T^{-1}JT$ will be a conformal involution as soon as k is sufficiently small. Since it has $2g+2$ fixpoints we may conclude that X_0 is hyperelliptic, and hence that H_g is closed.

5. Variation of the differentials

5.1. As in the preceding sections we shall denote local variables on the base surface (X_0, Σ_0) by z and variables on the surface (X, Σ) by ζ. The notation $\theta_i dz$ will refer to the differential on (X_0, Σ_0) with period 1 along a_i and period 0 along all a_k, $k \neq i$. The corresponding differential on (X, Σ), determined by the same conditions, is designated by $\theta_i^* d\zeta$.

Since the a-periods are the same, it follows by Riemann's bilinear relation that

$$\iint_{X_0} (\theta_i dz - \theta_i^* d\zeta) \wedge (\bar{\theta}_i d\bar{z} - \bar{\theta}_i^* d\bar{\zeta}) = 0.$$

By virtue of relation (1) this equation can be rewritten in the form

$$\iint_{X_0} \left| \theta_i - \theta_i^* \sqrt{\frac{f}{\phi}} \right|^2 dxdy = k^2 \iint_{X_0} |\theta_i^*|^2 \left| \frac{f}{\phi} \right| dxdy, \tag{4}$$

where $dxdy$ is the area element and the sign of $\sqrt{f/\phi}$ is uniquely determined.

If we use a square norm notation, (4) can be written as

$$\left\| \theta_i - \theta_i^* \sqrt{\frac{f}{\phi}} \right\| = k \left\| \theta_i^* \sqrt{\frac{f}{\phi}} \right\|.$$

We deduce that

$$\left\| \theta_i^* \sqrt{\frac{f}{\phi}} \right\| \leqq \frac{1}{1-k} \|\theta_i\| \tag{5}$$

and

$$\left\| \theta_i - \theta_i^* \sqrt{\frac{f}{\phi}} \right\| \leqq \frac{k}{1-k} \|\theta_i\|. \tag{6}$$

5.2. We have denoted the b_j-period of $\theta_i dz$ by τ_{ij}, and now we let τ_{ij}^* have the corresponding meaning for $\theta_i^* d\zeta$. It is a further immediate consequence of Riemann's relation that

$$\tau_{ij}^* - \tau_{ij} = \iint_{X_0} \theta_j dz \wedge (\theta_i^* d\zeta - \theta_i dz) = \iint_{X_0} \theta_j dz \wedge \theta_i^* d\zeta.$$

By use of (1) we can eliminate $d\zeta$ and obtain

$$\tau_{ij}^* - \tau_{ij} = k \iint\limits_{X_0} \theta_i^* \theta_j \sqrt{\frac{\bar{f}}{\phi}}\, dz \wedge d\bar{z}.$$

We rewrite this in the form

$$\tau_{ij}^* - \tau_{ij} = k \iint\limits_{X_0} \theta_i \theta_j \frac{\bar{f}}{|f|}\, dz \wedge d\bar{z} + \eta_{ij}, \tag{7}$$

where

$$\eta_{ij} = k \iint\limits_{X_0} \left(\theta_i^* \sqrt{\frac{\bar{f}}{\phi}} - \theta_i\right) \theta_j \frac{\bar{f}}{|f|}\, dz \wedge d\bar{z}.$$

Here $dz \wedge d\bar{z} = -2idxdy$, and by virtue of (6) we have thus

$$|\eta_{ij}| \leqq \frac{2k^2}{1-k} \|\theta_i\| \cdot \|\theta_j\|. \tag{8}$$

We wish to point out that the method we have used is an Hadamard variation. Similar formulas are found in two articles by Rauch [6, 7] whose work has many points of contact with ours.

6. Local mappings

6.1. If $g = 2$ we let X_0 denote an arbitrary point of M_g; if $g > 2$ we assume that X_0 is not hyperelliptic. In either case it is known that there are exactly $N = 3g-3$ linearly independent products $\theta_i\theta_j$. We shall denote by I the set of index-pairs (i, j) that corresponds to such a choice. We are going to show that there exists a neighborhood of X_0 which is mapped in a one-to-one manner by the functions $\tau_{ij}(X)$, $(i, j) \in I$.

We prove first that it is impossible to have $\tau_{ij}^* = \tau_{ij}$ for all $(i, j) \in I$ when k is sufficiently small. By (7) this assumption would imply

$$k \iint\limits_{X_0} \theta_i \theta_j \frac{\bar{f}}{|f|}\, dz \wedge d\bar{z} = -\eta_{ij}. \tag{9}$$

Since the $\theta_i\theta_j$ form a base for the quadratic differentials, it is possible to write

$$f = \sum_I c_{ij} \theta_i \theta_j. \tag{10}$$

From (9) we would therefore obtain

$$2k \iint\limits_{X_0} |f| dxdy \leqq \sum_I |c_{ij}|\, |\eta_{ij}|,$$

and by virtue of (8) we would have

$$\iint_{X_0} |f| dxdy \leqq \frac{k}{1-k} \sum_I |c_{ij}| \, \|\theta_i\| \, \|\theta_j\|.$$

It is clear that

$$\iint_{X_0} |f| \, dxdy$$

is a continuous function of the c_{ij}. Therefore it has a positive minimum when the coefficients are normalized, for instance, by the condition $\sum_I |c_{ij}| \, \|\theta_i\| \, \|\theta_j\| = 1$. We conclude that the equations $\tau^*_{ij} = \tau_{ij}$ cannot be fulfilled when k is smaller than a certain k_0.

6.2. When $g > 2$ we know that X remains non-hyperelliptic when $\bar{X}$ is in a sufficiently small neighborhood of $\bar{X}_0$. For such $\bar{X}$ the same conclusion as in 6.1 can thus be drawn, but it is important for us to verify that one can use the same index set I. Let us therefore assume that $\sum_I c_{ij}\theta^*_i\theta^*_j = 0$. On writing

$$\theta_i\theta_j - \theta^*_i\theta^*_j \frac{f}{\phi} = \theta_i\left(\theta_j - \theta^*_j\sqrt{\frac{f}{\phi}}\right) + \theta^*_j\sqrt{\frac{f}{\phi}}\left(\theta - \theta^*_i\sqrt{\frac{f}{\phi}}\right)$$

we obtain

$$\iint_{X_0} |\sum_I c_{ij}\theta_i\theta_j| dxdy \leqq \sum_I |c_{ij}| \left(\|\theta_i\| \left\| \theta_j - \theta^*_j\sqrt{\frac{f}{\phi}} \right\| + \left\| \theta^*_j\sqrt{\frac{f}{\phi}} \right\| \left\| \theta_i - \theta^*_i\sqrt{\frac{f}{\phi}} \right\| \right).$$

By use of the estimates (5) and (6) we have hence

$$\iint_{X_0} |\sum_I c_{ij}\theta_i\theta_j| dxdy \leqq \frac{k(2-k)}{(1-k)^2} \sum_I |c_{ij}| \, \|\theta_i\| \, \|\theta_j\|.$$

This is clearly impossible for small k, unless $\sum_I c_{ij}\theta_i\theta_j = 0$. In other words, if the $\theta_i\theta_j$ are linearly independent, so are the $\theta^*_i\theta^*_j$, provided that k is sufficiently small.

6.3. If we would possess a uniform lower bound for k_0, it would follow at once that the mapping is one-to-one in a neighborhood of $\bar{X}_0$. As a matter of fact, such a uniform bound could easily be found in the present circumstances. Later, however, we shall have to draw the same conclusion without use of uniformity, and for this reason we prefer to proceed without taking advantage of the uniform lower bound.

At this point it is convenient to shift to a vector notation. We let $\tau(\mathfrak{X})$ be the vector with components $\tau_{ij}(\mathfrak{X})$, $(i, j) \in I$. Similarly, η will be the vector with components η_{ij}. Finally, let $A(f)$ be the vector with components

$$(11) \qquad A_{ij}(f) = \iint_{X_0} \theta_i \theta_j \frac{\bar{f}}{|f|} \, dz \wedge d\bar{z}.$$

With these notations (7) can be written in the form

$$(12) \qquad \tau(\mathfrak{X}) - \tau(\mathfrak{X}_0) = kA(f) + \eta,$$

and the estimate (8) shows that $|\eta| = O(k^2)$. Let us also note that the vector norm $|A(f)|$ has a positive lower bound, independent of f.

We recall that the points $\mathfrak{X}$ are in one-to-one correspondence with the pairs (f, k), except for a positive factor in f. We show first that the mapping $(f, k) \to kA(f)$ is one-one, subject to the same condition. Suppose that $kA(f) = k_1A(g)$. In view of the representation (10) and the expressions (11) for the components A_{ij}, it follows at once that

$$(13) \qquad k \iint_{X_0} |f| dxdy = k_1 \iint_{X_0} \frac{f\bar{g}}{|g|} dxdy$$

as well as

$$(14) \qquad k_1 \iint_{X_0} |g| dxdy = k \iint_{X_0} \frac{g\bar{f}}{|f|} dxdy.$$

From (13) we obtain $k \leqq k_1$; and from (14), $k_1 \leqq k$. But with $k = k_1$, (13) can hold only so that f/g is positive. Since it is an analytic function it must reduce to a constant, and we have proved the one-to-oneness.

6.4. The mapping which takes $\mathfrak{X}$ into $kA(f)$ is evidently continuous, and we have just proved that it is one-one. Therefore, it is topological on every compact part of M_g, and in view of the simple effect that accompanies a change of k, it is even seen to be topological on all of M_g. The degree of a topological mapping is constantly either 1 or -1; we may number the coordinates so that it is 1.

We can choose $k_1 > 0$ so small that $|\eta| < |kA(f)|$ for $k = k_1$. It follows, by use of the generalized Rouché theorem, that the mapping $\mathfrak{X} \to \tau(\mathfrak{X})$, represented by (12), also has degree 1 at $\mathfrak{X}_0$. This result can be applied to any point in a neighborhood of $\mathfrak{X}_0$. Since $\tau(\mathfrak{X})$ takes its values at isolated points, we can further conclude that the algebraic sum of the mapping degrees at the points with a given image, sufficiently close to $\tau(\mathfrak{X}_0)$, is still equal to 1. This means that each value is taken once. Hence $\tau(\mathfrak{X})$ yields a topological mapping of a neighborhood of $\mathfrak{X}_0$.

7. The analytic structure

7.1. We are now ready to prove that the τ_{ij} generate a complex analytic structure on $M_g - H_g$ (or on M_g if $g = 2$). Since we have already shown that the τ_{ij} with $(i, j) \in I$ yield a topological mapping of a neighborhood V of $\bar{X}_0$, we know that any τ_{mn} can be written as a function of these τ_{ij} on the image of V. It remains to show the analyticity, and for this purpose we need only prove the existence of the partial derivatives at $\bar{X}_0$.

There exists a representation

$$\theta_m\theta_n = \sum_I e_{ij}\theta_i\theta_j.$$

By use of (7) we have thus

$$\tau^*_{mn} - \tau_{mn} = k\iint_{X_0} \theta_m\theta_n \frac{f}{|f|}\,dz \wedge d\bar{z} + \eta_{mn}$$

$$= \sum_I e_{ij}(\tau^*_{ij} - \tau_{ij}) + \eta_{mn} - \sum_I e_{ij}\eta_{ij},$$

and by (8) we obtain

$$\tau^*_{mn} - \tau_{mn} = \sum_I e_{ij}(\tau^*_{ij} - \tau_{ij}) + O(k^2), \tag{15}$$

where the estimate of the remainder is independent of f.

We know, further, that $|\tau^* - \tau|$ is greater than a positive constant times k. Therefore, (15) can be rewritten as

$$\tau^*_{mn} - \tau_{mn} = \sum_I e_{ij}(\tau^*_{ij} - \tau_{ij}) + O(\sum_I |\tau^*_{ij} - \tau_{ij}|^2). \tag{16}$$

This relation shows that τ_{mn} is an analytic function of the τ_{ij}, and moreover that the derivatives at the point $\bar{X}_0$ are given by

$$\frac{\partial\tau_{mn}}{\partial\tau_{ij}} = e_{ij}.$$

7.2. We prove now that the complex analytic structure which we have just introduced is unique, provided that we compare it with structures over the Teichmüller topology. Let $(z_1, \ldots, z_{3g-3})$ be a local base for a complex analytic structure on $M_g - H_g$ over the Teichmüller topology, and assume that the τ_{ij} are analytic functions of these z_k. We restrict all functions to a neighborhood V of a point $\bar{X}_0$. For a certain selection $(i, j) \in I$, we know that the mapping $z \rightarrow \tau$, defined on $z(V)$, is one-to-one. But then the Jacobian of the τ_{ij} with respect to the z_k cannot vanish at $z(\bar{X}_0)$ (Osgood, *Lehrbuch der Funktionentheorie*, II.1, p. 149). Therefore, the z_k can be expressed as analytic functions of the τ_{ij}, and the two structures are equivalent.

7.3. All that we have said about $M_g - H_g$ can be repeated for the components of H_g. The only difference is that the number of linearly independent products $\theta_i\theta_j$ is now $2g-1$. For each component of H_g there will exist a unique complex analytic structure over the Teichmüller topology which makes the functions τ_{ij} analytic. These structures are of complex dimension $2g-1$.

It should be noted that the Teichmüller topology on H_g coincides with the topology which is determined by the location of the branch-points in the standard representation of a hyperelliptic surface.

8. Some local estimates

8.1. In the following we consider a fixed differential θdz with simple zeros on the base surface X_0; in our applications X_0 will be hyperelliptic, but this is inessential. We choose a canonical system on X_0 and denote the corresponding point on M_g by $\mathfrak{X}_0$. On any $\mathfrak{X}$ there is then a unique differential $\theta^* d\zeta$ with the same a-periods as θdz. Our first task will be to show that θ^* has simple zeros as soon as $\mathfrak{X}$ is sufficiently close to $\mathfrak{X}_0$.

It is shown exactly as in 5.1 that

$$\left\| \theta - \theta^* \sqrt{\frac{f}{\phi}} \right\| \leqq \frac{k}{1-k} \|\theta\|. \tag{17}$$

However, for the conclusions that we have to make it will be necessary to replace this global inequality by one that holds at individual points.

One of the difficulties is that the parametric regions on X vary with f and k. In order to meet this difficulty it is convenient to pass to the universal covering surfaces of X_0 and X which we map conformally onto $|z| < 1$ and $|\zeta| < 1$. The Teichmüller mapping becomes a quasiconformal mapping $z \to \zeta$ of the unit disk onto itself, and it is known that such a mapping remains continuous on the boundary. For this reason we can normalize the mapping so that three boundary points have prescribed images, for instance, so that 1, i, and -1 remain fixed. With this agreement ζ, θ^*, and ϕ can be regarded as functions of z.

8.2. In the following we shall use the notation A_n for constant bounds that depend only on $\mathfrak{X}_0$ and are valid as soon as k is less than a certain k_n. Less explicitly, we write $O(k)$ for any quantity known to be $\leqq A_n k$ in absolute value; it is highly important that this limitation be independent of f and k, provided only that $k < k_n$.

We shall make constant use of

LEMMA 1. $|\zeta - z| \leqq A_1 k$.

For the proof we note that the quasiconformal mapping can be extended by symmetry to the whole plane without increasing the maximal dilatation.

The extremal quasiconformal mapping of the sphere which carries 1, i, $-i$, z into 1, i, $-i$, ζ is known, for its determination can be reduced to the corresponding problem for a pair of toruses. It is found that the Teichmüller distance equals the shortest non-euclidean distance between two values $\omega(z)$ and $\omega(\zeta)$, where ω is the function that maps the universal covering surface of the sphere, punctured at 1, i, $-i$, conformally on a disk. It can therefore be deduced that $|\zeta - z| = O(k)$ as soon as z stays away from the fixpoints. However, the restriction can be removed by applying the same result twice in succession, and we obtain, indeed, a universal inequality of the form $|\zeta - z| \leqq A_1 k$.

8.3. Each point on X_0 corresponds to infinitely many z, which are equivalent under the transformations of a Fuchsian group. It is possible to choose $R < 1$ so that the disk $|z| < R$ contains at least one point from each equivalence class. Next, we can determine $\rho < \frac{1}{4}(1 - R)$ so that no disk $|z - z_0| \leqq 2\rho$ with $|z_0| \leqq \frac{1}{2}(1 + R)$ contains more than one point from any equivalence class. The numbers R and ρ will be kept fixed.

LEMMA 2. $|\theta^*| \leqq A_2$ *for* $|z| \leqq \frac{1}{2}(1 + R)$.

We choose $k_2 = \min(k_1, \rho/2A_1)$ and assume henceforth that $k < k_2$. Let z_0 and ζ_0 be corresponding points with $|z_0| \leqq \frac{1}{2}(1 + R)$. The disk $|\zeta - \zeta_0| < \rho$ has its inverse image contained in $|z - z_0| < 2\rho$. Hence it contains no two equivalent points, and from

$$\theta^*(z_0)^2 = \frac{1}{\pi\rho^2} \iint_{|\zeta-\zeta_0|<\rho} \theta^{*2} d\xi d\eta$$

we conclude that

$$|\theta^*(z_0)|^2 \leqq \frac{1}{\pi\rho^2} \iint_{X_0} |\theta^*|^2 d\xi d\eta = \frac{1-k^2}{\pi\rho^2} \left\| \theta^* \sqrt{\frac{f}{\phi}} \right\|^2.$$

On combining with the estimate (5) we find that $|\theta^*(z_0)| \leqq A_2$ as claimed.

8.4. This time we assume that $|z_0| \leqq R$. Then $|\zeta - \zeta_0| < \rho$ implies $|z| < R + 2\rho < \frac{1}{2}(1 + R)$, so that Lemma 2 is applicable at all such points. The image of the disk $|z - z_0| \leqq \rho_1 = \rho - 2A_1 k$ is contained in $|\zeta - \zeta_0| \leqq \rho$ and contains the smaller disk $|\zeta - \zeta_0| \leqq \rho - 4A_1 k$. Therefore, the part of the larger disk which is not covered by the image has area $O(k)$. We have thus

$$(18) \qquad \begin{aligned} \theta^*(z_0) &= \frac{1}{\pi\rho^2} \iint_{|\zeta-\zeta_0|<\rho} \theta^* d\xi d\eta \\ &= \frac{1-k^2}{\pi\rho^2} \iint_{|z-z_0|<\rho_1} \theta^* \left|\frac{f}{\phi}\right| dx dy + O(k). \end{aligned}$$

We will use (18) to prove

LEMMA 3. $|\theta - \theta^*| \leqq A_3 k$ *for* $|z| \leqq R$.

We write

$$\iint_{|z-z_0|<\rho_1} \left(\theta - \theta^* \left|\frac{f}{\phi}\right|\right) dxdy = \iint_{|z-z_0|<\rho_1} \theta\left(1 - \sqrt{\frac{\bar{f}}{\phi}}\right) dxdy \tag{19}$$

$$+ \iint_{|z-z_0|<\rho_1} \left(\theta - \theta^* \sqrt{\frac{f}{\phi}}\right) \sqrt{\frac{\bar{f}}{\bar{\phi}}}\, dxdy.$$

The second term on the right can be estimated by means of the Schwarz inequality. We have indeed

$$\iint_{|z-z_0|<\rho_1} \left|\theta - \theta^* \sqrt{\frac{f}{\phi}}\right|^2 dxdy \leqq \left\|\theta - \theta^* \sqrt{\frac{f}{\phi}}\right\|^2 = O(k^2),$$

and

$$\iint_{|z-z_0|<\rho_1} \left|\frac{f}{\phi}\right| dxdy \leqq \frac{1}{1-k^2} \iint_{|\zeta-\zeta_0|<\rho} d\xi d\eta = \frac{\pi\rho^2}{1-k^2}.$$

We conclude that the second term is of order $O(k)$.

As for the first term we note that $\sqrt{\frac{\bar{f}}{\phi}}\, dz \wedge d\bar{z} = dz \wedge d\bar{\zeta}$. Hence

$$\iint_{|z-z_0|<\rho_1} \theta\left(1 - \sqrt{\frac{\bar{f}}{\phi}}\right) dz \wedge d\bar{z} = \iint_{|z-z_0|<\rho_1} \theta dz \wedge d(\bar{z} - \bar{\zeta})$$

$$= - \int_{|z-z_0|=\rho_1} (\bar{z} - \bar{\zeta})\theta dz,$$

and by Lemma 1 this term is likewise of order $O(k)$. On combining (18) and (19) we find

$$\theta^*(z_0) = (1 - k^2) \frac{\rho_1^2}{\rho^2} \theta(z_0) + O(k)$$

and the lemma follows since $\rho_1/\rho = 1 + O(k)$.

8.5. Suppose that $\theta(z_0) = 0$, $|z_0| < R$, and choose $r < R - |z_0|$ so small that θ has no other zeros inside the circle $|z - z_0| = r$. Then $|\theta|$ has a positive minimum on the circle, and by Lemma 3 we shall have $|\theta - \theta^*| < |\theta|$ on the same circle as soon as k is sufficiently small. We conclude by Rouché's theorem that θ^* has exactly one zero z_0^* inside the circle. A further application of Lemma 3 gives $|\theta(z_0^*)| \leqq A_3 k$. Since z_0 is a simple zero this implies $|z_0^* - z_0| = O(k)$. We have proved:

LEMMA 4. *The zeros of θ and θ^* can be numbered so that they have representatives z_i, z_i^* in $|z| < R$ which satisfy $|z_i^* - z_i| \leqq A_4 k$.*

9. The variational formula for τ_{g+j}

9.1. Given θdz on $\bar{X}_0$, we have defined $\theta^* d\zeta$ as the differential on $\bar{X}$ with the same a-periods. The b-periods of $\theta^* d\zeta$ will be denoted by $\tau_i(\bar{X})$, $i = 1, \ldots, g$. They are well defined functions of $\bar{X}$, and for brevity we write $\tau_i(\bar{X}) = \tau_i^*$, $\tau_i(\bar{X}_0) = \tau_i$. The variational formulas expressed by (5) and (6) remain valid and can be written in the form

$$\tau_i^* - \tau_i = k \iint_{X_0} \theta\theta_i \frac{f}{|f|} \, dz \wedge d\bar{z} + O(k^2). \tag{20}$$

9.2. We determine the zeros z_i, z_i^* in accordance with Lemma 4. Let σ_j be an arbitrary but fixed path from z_1 to z_{1+j}, $j = 1, \ldots, 2g-3$, within $|z| < 1$. To this path we add the straight line segments from z_i^* to z_1 and from z_{1+j} to z_{i+j}^*; the resulting path is denoted by σ_j^*. We use the same notation for its image under the mapping $z \to \zeta$, and even for the corresponding path on X. The functions

$$\tau_{g+j}(\bar{X}) = \int_{\sigma_j^*} \theta^* d\zeta$$

are defined as soon as $\bar{X}$ is in the neighborhood of $\bar{X}_0$ determined by $k < k_4$. We are going to show that $\tau_1, \ldots, \tau_{3g-3}$ determine an analytic structure.

9.3. The cycles a_i, b_i on X_0 can be chosen so that they do not pass through the points z_j. For such a choice let $\omega_j dz$ be the differential of third kind on X_0 which has zero a-periods and simple poles of residue 1 at z_{1+j} and -1 at z_1.

By contour integration we obtain

$$\iint_X \omega_j dz \wedge \theta^* d\zeta = \iint_X \omega_j dz \wedge (\theta^* d\zeta - \theta dz) = 2\pi i \int_{\sigma_j} \theta dz - \theta^* d\zeta \tag{21}$$

On introducing the notations τ_{g+j}, τ_{g+j}^* we have

$$\int_{\sigma_j} \theta dz - \theta^* d\zeta = \tau_{g+j} - \tau_{g+j}^* + \int_{z_1}^{z_1^*} \theta^* d\zeta - \int_{z_{1+i}}^{z_{1+i}^*} \theta^* d\zeta.$$

The integrals on the right can be evaluated along straight line segments in the ζ-plane. The image of the disk $|z| \leqq \frac{1}{2}(1 + R)$ contains $|\zeta| \leqq \frac{1}{2}(1 + R) - A_1 k$, while $|\zeta| \leqq R + A_1 k$ on the line from ζ_1 to ζ_1^*, say. If we assume that $A_1 k < \frac{1}{8}(1 - R)$, it follows that each point on the line is the center of a disk of radius $\frac{1}{4}(1 - R)$ in which the estimate $|\theta^*| \leqq A_2$ is valid. We conclude by Cauchy's formula that $d\theta^*/d\zeta$ is bounded. Since θ^* vanishes at ζ_1^*, we see that $\theta^* = O(|\zeta_1 - \zeta_1^*|) = O(k)$ on the line, and finally that

$$\int_{\zeta_1}^{\zeta_1^*} \theta^* d\zeta = O(k^2).$$

The same reasoning applies to the integral from z_{1+j} to z_{1+j}^*. Hence (22) yields

$$\tau_{g+j}^* - \tau_{g+j} = -\frac{1}{2\eta i} \iint_X \omega_j dz \wedge \theta^* d\zeta + O(k^2)$$

or, when $d\zeta$ is eliminated,

(22)
$$\tau_{g+j}^* - \tau_{g+j} = \frac{k}{\pi} \iint_{X_0} \omega_j \theta^* \sqrt{\frac{\bar{f}}{\phi}} \, dxdy + O(k^2).$$

9.4. It will now be necessary to estimate

(23)
$$\iint_{X_0} \omega_j \left(\theta^* \sqrt{\frac{\bar{f}}{\phi}} - \theta \sqrt{\frac{\bar{f}}{f}} \right) dxdy.$$

We begin by examining the parts of the integral that correspond to small disks $|z - z_1| < r$ and $|z - z_{1+j}| < r$. Since θ has a zero which compensates the pole z_1 of ω_j, it is immediately clear that

$$\iint_{|z-z_1|<r} \omega_j \theta \sqrt{\frac{\bar{f}}{f}} \, dxdy = O(r^2).$$

Next we write

$$\iint_{|z-z_1|<r} \omega_j \theta^* \sqrt{\frac{\bar{f}}{\phi}} \, dxdy = \theta^*(z_1) \iint_{|z-z_1|<r} \omega_j \sqrt{\frac{\bar{f}}{\phi}} \, dxdy$$

$$+ \iint_{|z-z_1|<r} \omega_j(\theta^* - \theta^*(z_1)) \sqrt{\frac{\bar{f}}{\phi}} \, dxdy.$$

In the first integral we use integration by parts to obtain

$$\iint_{|z-z_1|<r} \omega_j \sqrt{\frac{\bar{f}}{\phi}} \, dz \wedge d\bar{z} = \frac{1}{k} \iint_{|z-z_1|<r} \omega_j dz \wedge d\zeta$$

$$= -\frac{1}{k} \left[\int_{|z-z_1|=r} (\zeta - z)\omega_j dz - (\zeta_1 - z_1) 2\pi i \right].$$

It follows by Lemma 1 that this integral is bounded. As for the factor $\theta^*(z_1)$, we have $|\theta^*(z_1)| = |\theta^*(z_1) - \theta(z_1)| = O(k)$ by Lemma 3. Hence the contribution of the first term is of order $O(k)$.

In the second term we find easily that $|\theta^* - \theta^*(z_1)| = O(|\zeta - \zeta_1|)$, by the same reasoning as in 9.3. To continue the estimate we need a result from the general theory of quasiconformal mappings. As soon as the dilatation is $\leq K$, it is known that $|\zeta - \zeta_1| < A_5|z - z_1|^{1/K}$ with a constant that does not depend on the mapping. If we assume that $k < 1/2$, we have thus $|\theta^* - \theta^*(z_1)| = O(|z - z_1|^{1/3})$.

This estimate enables us to apply Schwarz' inequality. We find

$$\left| \iint_{|z-z_1|<r} \omega_j(\theta^* - \theta^*(z_1)) \sqrt{\frac{\bar f}{\phi}}\, dxdy \right|^2$$

$$\leq \iint_{|z-z_1|<r} |\omega_j|^2 |\theta^* - \theta^*(z_1)|^2 dxdy \cdot \iint_{|z-z_1|<r} \left| \frac{f}{\phi} \right| dxdy.$$

The first factor on the right is of order $O(r^{2/3})$. The second factor is trivially bounded, for it is essentially the area of the image of $|z - z_1| < r$ in the ζ-plane (the sharper estimate $O(r^{2/3})$ would not lead to any essential improvement in the final result).

The same reasoning applies to the disk $|z - z_{s+j}| < r$. On the remaining part of X_0 we have $\omega_j = O(1/r)$. We combine this with the trivial estimate

$$\iint_{X_0} \left| \theta^* \sqrt{\frac{\bar f}{\phi}} - \theta \sqrt{\frac{\bar f}{f}} \right| dxdy \leq \sqrt{\pi}\, R \left\| \theta^* \sqrt{\frac{\bar f}{\phi}} - \theta \right\| = O(k).$$

When the results are collected we find that the integral (23) is of order $O(k) + O(r^{2/3}) + O(k/r)$. On choosing $r = k^{3/5}$ it follows that the order is $O(k^{2/5})$, and hence (22) leads to the variational formula

$$(24) \qquad \tau^*_{g+j} - \tau_{g+j} = \frac{k}{\pi} \iint_{X_0} \omega_j \theta \frac{f}{|f|}\, dxdy + O(k^{7/5}).$$

We emphasize once more that the estimate of the remainder is uniform as long as X_0 remains fixed. We have not pushed the estimate far enough to permit X_0 to vary. To do so would require us to study the variation of ω_j, and this seems to be much more difficult.

10. Local mapping by the τ_t

10.1. It is evident that $\theta_1\theta$, $\theta_2\theta$, ..., $\theta_g\theta$, $\omega_1\theta$, ..., $\omega_{2g-3}\theta$ are linearly independent. Since their number coincides with the dimension of the space of quadratic differentials they form a base. Thus, if $f dz^2$ is an everywhere regular quadratic differential we can write

$$(25) \qquad f = \sum_1^g c_i\theta_i\theta + \sum_1^{2g-3} c_j\omega_j\theta.$$

We are now in a position to show that the mapping by the vector function $\tau(\bar{X}) = (\tau_1(\bar{X}), \dots, \tau_{3g-3}(\bar{X}))$ is a homeomorphism in a suitably restricted neighborhood of $\bar{X}_0$. It turns out that we are almost in the same situation that prevailed in **6**. For this reason the description of the necessary steps will be very brief.

We show first that $\tau^* \neq \tau$ for sufficiently small k. By (20) and (24) the case of equality would imply

$$\iint_{X_0} \theta\theta_i \frac{\bar{f}}{|f|}\, dxdy = O(k)$$

$$\iint_{X_0} \theta\omega_j \frac{\bar{f}}{|f|}\, dxdy = O(k^{2/5}).$$

When these equations are multiplied by c_i, c_j and added it would follow that

$$\iint_{X_0} |f|\, dxdy \leqq A_6 k^{2/5} (\Sigma\, |c_i| + \Sigma\, |c_j|),$$

and this is clearly impossible for small k. Hence, there exists a neighborhood of $\bar{X}_0$ in which $\tau(\bar{X}) \neq \tau(\bar{X}_0)$.

10.2. In order to duplicate the reasoning in 6.3, we introduce the vector $A(f)$ whose coordinates are

$$A_i(f) = \iint_{X_0} \theta\theta_i \frac{\bar{f}}{|f|}\, dz \wedge d\bar{z}$$

for $i = 1, \dots, g$, and

$$A_{g+j}(f) = \frac{1}{\pi} \iint_{X_0} \omega_j \theta \frac{\bar{f}}{|f|}\, dxdy$$

for $j = 1, \dots, 2g-3$. The equations (20) and (24) can be collected to

$$\tau(\bar{X}) - \tau(\bar{X}_0) = kA(f) + O(k^{7/5}).$$

It is proved as in 6.3 that the mapping $(f, k) \to kA(f)$ is one-one, provided that f is normalized. The degree of the mapping $\bar{X} \to kA(f)$ is therefore constantly 1, and by application of the generalized Rouché theorem it follows that the local and total algebraic degrees of the mapping $\bar{X} \to \tau(\bar{X})$ are 1 for all values sufficiently close to $\tau(\bar{X}_0)$. For this reason the mapping is a local homeomorphism.

10.3. The reasoning in 7.1 can be duplicated to show that the functions τ_j define an analytic structure. In the present case we have to compare two vectors τ and τ' which are defined in overlapping neighborhood V and V'. We recall that the coordinates τ_j do not depend on the base point

$\bar{X}_0$, except for the choice of a particular differential θ on X_0. However, if we fix θ at one point of V, it is automatically fixed for all points in V. Therefore, when we compare τ and τ', we can use the same base points $\bar{X}_0 \in V \cap V'$, but we may have to consider different differentials θ and θ'.

Let us write $h_j = \theta\theta_j$ for $j = 1, \ldots, g$ and $h_{g+j} = (2\pi i)^{-1}\theta\omega_j$, $j = 1, \ldots, 2g-3$. With a similar meaning of h'_j we have

$$\tau_j(\bar{X}) - \tau_j(\bar{X}_0) = k \iint\limits_{X_0} h_j \frac{\bar{f}}{|f|}\, dxdy + O(k^{7/5}),$$

and

$$\tau'_j(\bar{X}) - \tau'_j(\bar{X}_0) = k \iint\limits_{X_0} h'_j \frac{\bar{f}}{|f|}\, dxdy + O(k^{7/5}).$$

But the $h_j dz^2$ are a basis for the quadratic differentials on X_0. Therefore, h'_j can be expressed linearly as

$$h'_j = \sum_k e_{jk} h_k,$$

and we obtain

$$\tau'_j(\bar{X}) - \tau'_j(\bar{X}_0) = \sum_k e_{jk}(\tau_k(\bar{X}) - \tau_k(\bar{X}_0)) + O(k^{7/5}).$$

Here k is of magnitude $O(|\tau(X) - \tau(X_0)|)$. We conclude that each τ'_j is an analytic function of the τ_k.

Precisely the same argument shows that each τ_{mn} is an analytic function of the τ_j. We have thus proved the existence of an analytic structure which makes the elements of the Riemann matrix analytic.

11. The uniqueness

11.1. It remains to show that the analytic structure is unique. We have already proved that the structure is unique on $M_g - H_g$, so we need only consider the neighborhood of a point $\bar{X}_0 \in H_g$. We assume, therefore, that $z_1, \ldots, z_{3g-3}$ is a basis for an analytic structure in a neighborhood V of $\bar{X}_0$, and that the τ_{ij} are analytic with respect to this basis.

We may suppose that the functions $\tau_1, \ldots, \tau_{3g-3}$ which we have introduced are all defined in V. The first g, $\tau_1, \ldots, \tau_g$, are linear combinations of the τ_{ij}, and thus are known to be analytic. The remaining ones are known to be analytic on $V - H_g$ and continuous on V. We shall invoke a classical theorem which shows that they are then automatically analytic. The theorem in question states that the singularities of an analytic function are removable if there exists an analytic function which vanishes on the exceptional set without being identically zero.

We proved in 7.3. that the τ_{ij} generate a $(2g-1)$-dimensional complex analytic structure on each component of H_g. Therefore, if V is sufficiently small, we can select a set I of $2g-1$ double indices (i, j) such that every τ_{mn} can be expressed as an analytic function of these τ_{ij} on $V \cap H_g$. Let the explicit expressions be $\tau_{mn} = f_{mn}(\tau_{ij})$. The functions $f_{mn}(\tau_{ij})$ remain analytic on all of V, provided that V is small enough, but if $g > 2$ the equations $\tau_{mn} = f_{mn}(\tau_{ij})$ cannot all be identically fulfilled on V, for then the complex dimension of V would be less than $3g-3$. Therefore, there exists a function $F_{mn} = \tau_{mn} - f_{mn}(\tau_{ij})$ which is analytic in $z_1, \ldots, z_{3g-3}$ on V, vanishes on $V - H_g$, and does not vanish identically. We conclude that $\tau_{g+1}, \ldots, \tau_{3g-3}$ are analytic functions of $z_1, \ldots, z_{3g-3}$.

Finally, since the mapping $z \rightarrow \tau$ is locally one-one, the Jacobian cannot vanish, and z can be expressed analytically through τ. It is proved that the analytic structure is unique.

Incidentally, we have also shown that the components of H_g are singularity-free analytic submanifolds of M_g.

REFERENCES

[1] L. AHLFORS, On quasiconformal mappings, *Journ. d'Analyse Math.* vol. III, 1 (1953/54).

[2] J. JENKINS, Some new canonical mappings for multiply-connected domains, *Ann. of Math.* **65** (1957).

[3] O. TEICHMÜLLER, Extremale quasikonforme Abbildungen und quadratische Differentiale, *Preuss. Akad.* **22** (1940).

[4] ———, Bestimmung der extremalen quasikonformen Abbildungen bei geschlossenen orientierten Riemannschen Flächen, *Preuss. Akad.* **4** (1943).

[5] ———, Veränderliche Riemannsche Flächen, *Deutsche Math.* **7** (1944).

Some Remarks on Perturbation of Structure

Donald C. Spencer
PRINCETON UNIVERSITY

THE basic technique used by Kodaira and Spencer in their paper [6] to describe the deformation of complex analytic structure is applied to other structures, in particular to real foliate structure (see, e.g., Reeb [7]) and to structures (real and complex) which are defined by simple infinite pseudo-groups in the sense of E. Cartan [1]. Comparison is made with the treatment of complex analytic structure described in [6]; this provides insight into the mechanism used and elucidates the rôle played by the theory of harmonic forms.

A systematic treatment of the deformation of real and complex multi-foliate structures will be published elsewhere by K. Kodaira and the author. The present paper contains a few examples, selected (from unpublished work of Kodaira and the author) to illustrate some of the techniques which are applicable to deformation of structure.

1. Definition of structure

Let $\mathscr{T}$ be the category of topological spaces and continuous maps, and let $\mathscr{M}$ be a subcategory of $\mathscr{T}$ satisfying the conditions that any open subset of an object of $\mathscr{M}$ is an object of $\mathscr{M}$ and that the restriction of any map of $\mathscr{M}$ to an open subset is also a map of $\mathscr{M}$. The maps of $\mathscr{M}$ will be called regular maps, and a map will be said to be biregular if it is a regular homeomorphism together with its inverse. We remark that $\mathscr{M}$ is a local category in the sense of Cartan-Eilenberg [2] for which the functor L : $\mathscr{M} \rightarrow \mathscr{T}$ (in this case inclusion) is faithful. However, we do not require a knowledge of Cartan-Eilenberg's definitions, but shall give a brief discussion of structure along lines formulated by Dr. H. K. Nickerson and the author.

Let X be a topological space. For each open set $U \subset X$, let $S(U) = \{\varphi \mid \varphi$ a homeomorphism from U onto an object of $\mathscr{M}\}$. If $U' \subset U''$, we have $S(U'') \rightarrow S(U')$ where $\varphi \in S(U'')$ goes into $\varphi \mid U'$. Next, let

$$G(U) = \{(\varphi, \psi) \mid \varphi, \psi \in S(U), \varphi \circ \psi^{-1} \text{ and } \psi \circ \varphi^{-1} \in \mathscr{M}\}.$$

We have the restriction map $G(U'') \to G(U')$ where (φ, ψ) goes into $(\varphi \mid U', \psi \mid U')$. An element of $G(U)$ is an identity if it is of the form (φ, φ), $\varphi \in S(U)$, and the product $(\varphi_1, \psi_1) \circ (\varphi_2, \psi_2)$ is defined in $G(U)$ if and only if $\psi_1 = \varphi_2$ in which case $(\varphi_1, \psi_1) \circ (\varphi_2, \psi_2) = (\varphi_1, \psi_2)$. The product clearly commutes with restriction so the restriction map is a homomorphism in the sense of groupoids.

The assignments $U \to G(U)$, $U \to S(U)$ define presheaves on X which induce sheaves on X to be denoted by S, G respectively. We shall say that $\varphi, \psi \in S(U)$ are equivalent if $(\varphi, \psi) \in G(U)$, and we shall denote the equivalence classes of $S(U)$ modulo $G(U)$ by $S(U)/G(U)$. Then $U \to S(U)/G(U)$ induces a sheaf S/G over X. We have: $H^0(X, S/G) \cong H^1(X, G)$.

DEFINITION 1.1. *An $\mathscr{M}$-structure on X is an element of $H^0(X, S/G)$, i.e. a section of S/G over X.*

REMARK: Each object M of $\mathscr{M}$ has a canonical $\mathscr{M}$-structure assigned by the identity map $\iota : M \to M$.

DEFINITION 1.2. *Given two categories $\mathscr{M}$, $\mathscr{N}$, we shall say that an $\mathscr{M}$-structure lies over an $\mathscr{N}$-structure (or that an $\mathscr{N}$-structure underlies an $\mathscr{M}$-structure) if $\mathscr{M}$ is a subcategory of $\mathscr{N}$.*

Thus every $\mathscr{M}$-structure has an underlying topological structure ($\mathscr{T}$-structure).

Next, let $\mathbf{K}$ stand either for the field $\mathbf{R}$ of real numbers or for the field $\mathbf{C}$ of complex numbers and, for each $X \in \mathscr{T}$, let $\mathscr{C}(X) = \mathscr{C}(X, \mathbf{K})$ denote the sheaf of germs of continuous functions from X to $\mathbf{K}$. Given a map $f : X \to Y$ of $\mathscr{T}$, there is a natural map $\mathscr{C}(f) : \mathscr{C}(Y) \to \mathscr{C}(X)$.

DEFINITION 1.3. *We say that a category $\mathscr{M}$ possesses regular functions (relative to continuous functions) if there is a map $\mathscr{R}$ from the objects of $\mathscr{M}$ into sheaves of rings of functions satisfying the following conditions:*

(i) *For $M \in \mathscr{M}$, $\mathbf{K} \subseteq \mathscr{R}(M) \subseteq \mathscr{C}(M)$ where $\mathbf{K}$ stands here for the constant sheaf over M.*

(ii) *A map f of $\mathscr{T}$ from an object M of $\mathscr{M}$ into an object N of $\mathscr{M}$ is regular if and only if $\mathscr{C}(f)(\mathscr{R}(N)) \subseteq \mathscr{R}(M)$, i.e. if the restriction of $\mathscr{C}(f)$ to $\mathscr{R}(N)$ is a map $\mathscr{R}(f) : \mathscr{R}(N) \to \mathscr{R}(M)$.*

(iii) *Given functors $\mathscr{S}$, $\mathscr{S}'$ satisfying* (i) *and* (ii), *we write $\mathscr{S}' < \mathscr{S}$ if and only if $\mathscr{S}'(M) \subseteq \mathscr{S}(M)$ for every object $M \in \mathscr{M}$; then $\mathscr{S} < \mathscr{R}$ implies that $\mathscr{S}$ coincides with $\mathscr{R}$, i.e. $\mathscr{R}$ is minimal with respect to* (i) *and* (ii).

If $\mathscr{M}$ is a category with regular functions, any $\mathscr{M}$-structure on a topological space X induces over X a sheaf which will be called the *sheaf of regular functions* on X. Hence we speak of "structures with regular functions", i.e. structures defined by categories with regular functions. We list here three classical structures of which only the first and third have regular functions (relative to continuous functions):

(i) *Differentiable structure.* $\mathscr{M}$ is the category whose objects are subdomains of $\mathbf{R}^n$ (real n-space) and whose maps are all differentiable maps of one subdomain of $\mathbf{R}^n$ onto another where "differentiable" here and elsewhere means "differentiable of class C^∞". The regular functions are all differentiable functions.

(ii) *Foliate structure* (*real*). The objects of $\mathscr{M}$ are the subdomains of $\mathbf{R}^n$ and the maps are the differentiable maps whose jacobians are matrices of the form

$$\begin{pmatrix} A & B \\ 0 & C \end{pmatrix} \tag{1.1}$$

where A is a $(p \times p)$-matrix, C is a $(q \times q)$-matrix where $p + q = n$, and B is a matrix of p rows and q columns. It is convenient to distinguish the last q coordinates by writing $x^1, \ldots, x^p, y^1, \ldots, y^q$ for the coordinates of $\mathbf{R}^n$.

(iii) *Complex analytic structure.* The objects are subdomains of $\mathbf{C}^n$ (complex n-space) and the maps are the holomorphic maps. The regular functions are the holomorphic functions.

We shall be concerned solely with structures which lie over differentiable structure in the sense of Definition 1.2, and by "structure" we shall henceforth mean a structure with an underlying differentiable structure. To such structures the methods of differential geometry are applicable. We suppose that all spaces are *paracompact.* A space which carries a structure will then have the underlying structure of a *differentiable manifold*, and we shall use the letter V to denote a space with a structure, the letter X to denote the underlying differentiable manifold of V.

The regular maps of foliate structure (example (ii) above) are characterized among all *differentiable* maps by means of a subsheaf of the sheaf of differentiable functions, namely the subsheaf of functions which depend only on $y^1, \ldots, y^q$. Since all maps are henceforth assumed to be differentiable, we shall call these functions "regular" although they do not satisfy the above conditions for regularity relative to *continuous* maps.

Finally, let X be a differentiable manifold of dimension n, and let F be a subgroup of the structure group $G = GL(n, \mathbf{R})$ of the tangent bundle of X. Denote by P the sheaf of germs of differentiable sections of the associated principal bundle of the tangent bundle of X, and call two sections equivalent if one is the transform of the other by an element of F. We thus obtain the sheaf P/F of equivalence classes of P modulo F, and an F-structure on X is defined by an element of $H^0(X, P/F)$. For example, almost-foliate structure is the F-structure defined by the subgroup of G composed of matrices (1.1); foliate structure is *integrable* almost-foliate structure. We note that an $O(n)$-structure is a Riemannian metric. It is clear that an F-structure is a special $\mathscr{M}$-structure in the sense of Definition 1.1.

2. Deformation of structure; mixed structure

We now introduce the notion of a differentiable family of $\mathcal{M}$-structures in the same way as this was done in [6] for the case of complex analytic structure. Namely, let B be a *connected* differentiable manifold, and let $(\mathcal{V}, \varpi, B)$ be a differentiable fibre bundle with base space B, total space $\mathcal{V}$, projection map $\varpi : \mathcal{V} \to B$, and with fibre a (connected) differentiable manifold X. Let $\mathcal{M}$ be a category of the type described at the beginning of Section 1 which is a subcategory of the category for differentiable structure, and suppose that each fibre $\varpi^{-1}(t)$, $t \in B$, carries a structure of $\mathcal{M}$-manifold V_t whose underlying differentiable structure is compatible with that of $\mathcal{V}$. We write $V_t = \varpi^{-1}(t)$, $t \in B$.

DEFINITION 2.1. *We say that $\mathcal{V} \to B$ is a differentiable family of $\mathcal{M}$-structures (or $\mathcal{M}$-manifolds) if each point of $\mathcal{V}$ has a neighborhood $\mathcal{U}$ together with a biregular differentiable map $h : \mathcal{U} \to M \times \varpi(\mathcal{U})$ of $\mathcal{U}$ onto the product of an object M of $\mathcal{M}$ with $\varpi(\mathcal{U})$ whose restriction to $\mathcal{U} \cap V_t$, $t \in \varpi(\mathcal{U})$, is a biregular map in the sense of $\mathcal{M}$ of $\mathcal{U} \cap V_t$ into $M \times t$.*

We shall always assume that both $\mathcal{V}$ and B are orientable.

Let $V_o = \varpi^{-1}(o)$ be the fibre of $\mathcal{V}$ over a particular point $o \in B$; then $V_t = \varpi^{-1}(t)$, $t \in B$, may be regarded as a deformation of the $\mathcal{M}$-structure of V_o.

We remark that a differentiable family $\mathcal{V}$ of $\mathcal{M}$-structures is, in particular, a foliate manifold whose leaves ("sheets") are $\mathcal{M}$-manifolds forming the fibres of a differentiable fibre bundle $\mathcal{V} \to B$.

A differentiable family of $\mathcal{M}$-structures is an example of a "mixed structure". A more general type of mixed structure might be defined as follows. Let $\mathcal{M}$, $\mathcal{N}$ be two categories each of which is a subcategory of the category for differentiable structure, and let $\mathcal{L}$ be the category whose objects are products $M \times N$ where $M \in \mathcal{M}$, $N \in \mathcal{N}$. The regular maps are chosen to be those differentiable maps whose restrictions to $M \times \mathfrak{q}$, $\mathfrak{q} \in N$, or to $\mathfrak{p} \times N$, $\mathfrak{p} \in M$, are regular in the senses of $\mathcal{M}$, $\mathcal{N}$ respectively. However, the mixed structure of a differentiable family in the sense of Definition 2.1 has the properties desired for defining "deformation of structure".

Consider now the special case of a differentiable family $\mathcal{V} \to B$ of foliate structures (compare example (ii) of Section 1), and let $\mathfrak{U} = \{\mathcal{U}_i\}$ be a locally finite covering of $\mathcal{V}$ by open sets $\mathcal{U}_i$ with associated maps $h_i : \mathcal{U}_i \to M_i \times \varpi(\mathcal{U}_i)$ as described in Definition 1.4. If we refer h_i to the local coordinates, we have

$$h_i(\mathfrak{p}) = (x_i^1(\mathfrak{p}), \ldots, x_i^p(\mathfrak{p}), y_i^1(\mathfrak{p}), \ldots, y_i^q(\mathfrak{p}), t),$$
$$t = \varpi(\mathfrak{p}), \ \mathfrak{p} \in \mathcal{U}_i.$$

We may also express the point $t \in \varpi(\mathcal{U}_i)$ in terms of local differentiable coordinates on B, namely $t = (t_i^1, \ldots, t_i^m)$, $m = \dim B$. We denote the tangent bundle of $\mathscr{V}$ by $\mathfrak{E}$; its transition functions are the matrices

$$(2.1)\qquad \begin{pmatrix} (\partial x_i^\alpha/\partial x_k^\beta) & (\partial x_i^\alpha/\partial y_k^\beta) & (\partial x_i^\alpha/\partial t_k^\nu) \\ 0 & (\partial y_i^\alpha/\partial y_k^\beta) & (\partial y_i^\alpha/\partial t_k^\nu) \\ & 0 & (\partial t_i^\mu/\partial t_k^\nu) \end{pmatrix}$$

which operate, on the left, on the components $(a_k^1, \ldots, a_k^p;\ b_k^1, \ldots, b_k^q;\ c_k^1, \ldots, c_k^m)$ of a tangent vector

$$(2.2)\qquad \textstyle\sum_\beta a_k^\beta \dfrac{\partial}{\partial x_k^\beta} + \sum_\beta b_k^\beta \dfrac{\partial}{\partial y_k^\beta} + \sum_\nu c_k^\nu \dfrac{\partial}{\partial t_k^\nu}.$$

Denote by $GL(r, \mathbf{R};\ s, \mathbf{R})$ the subgroup of $GL(r+s, \mathbf{R})$ which is composed of matrices of the form

$$(2.3)\qquad \begin{pmatrix} A & B \\ 0 & C \end{pmatrix}$$

where $A \in GL(r, \mathbf{R})$, $C \in GL(s, \mathbf{R})$, and where B is an arbitrary matrix of r rows and s columns. Given a subgroup F of $GL(r, \mathbf{R})$, let $GL(F;\ s, \mathbf{R})$ be the subgroup of $GL(r, \mathbf{R};\ s, \mathbf{R})$ composed of matrices (2.3) where A is required to belong to F. Then $\mathfrak{E}$ has structure group $GL(GL(p, \mathbf{R};\ q, \mathbf{R});\ m, \mathbf{R})$.

We remark that we can give an alternative definition of a differentiable family $\mathscr{V} \to B$ of foliate structures, namely:

Definition 2.2. *A differentiable family $\mathscr{V} \to B$ of foliate structures is a differentiable fibre bundle $(\mathscr{V}, \varpi, B)$ together with a differentiable reduction of the structure group $GL(n, \mathbf{R};\ m, \mathbf{R})$ of the tangent bundle of $\mathscr{V}$ to the subgroup $GL(GL(p, \mathbf{R};\ q, \mathbf{R});\ m, \mathbf{R})$, $p + q = n$, which imparts to each fibre a foliate structure (i.e. which imparts to each fibre an integrable almost-foliate structure).*

Definition 2.2 may be proved to be equivalent to Definition 2.1 (with the category for foliate structure), but the proof is tedious and will be omitted here (Definition 2.2 will not be used in the sequel).

Finally, we denote by $\mathfrak{F}$ the sub-bundle of $\mathfrak{E}$ composed of the vectors

$$(2.4)\qquad \textstyle\sum_\beta a_k^\beta \dfrac{\partial}{\partial x_k^\beta} + \sum_\beta b_k^\beta \dfrac{\partial}{\partial y_k^\beta}$$

along the fibres of $\mathscr{V}$, and we have the exact sequence

$$(2.5)\qquad 0 \to \mathfrak{F} \to \mathfrak{E} \to \mathfrak{E}/\mathfrak{F} \to 0.$$

We call (2.5) the *fundamental sequence of bundles* for the family $\mathscr{V} \to B$ of foliate structures.

The above fundamental sequence is entirely analogous to the corresponding fundamental sequence for complex analytic structure which is described in [6]. In the case of complex analytic structure replace x_k^α by z_k^α, y_k^α by $\bar{z}_k^\alpha(\alpha = 1,2, \ldots, n)$; then $\mathfrak{E}$ is the bundle of tangent vectors

$$\sum_\beta a_k^\beta \frac{\partial}{\partial z_k^\beta} + \sum_\nu c_k^\nu \frac{\partial}{\partial t_k^\nu} \tag{2.6}$$

and $\mathfrak{F}$ is the sub-bundle of vectors $\Sigma\, a_k^\beta(\partial/\partial z_k^\beta)$ along the fibres of $\mathscr{V}$ whose restriction to each fibre is the holomorphic tangent bundle of that fibre.

3. Fundamental sheaves; exponentiation

Let $\mathscr{V} \to B$ be a differentiable family of foliate structures, and denote by $\mathfrak{D}$ the sheaf of germs of (real-valued) differentiable functions on $\mathscr{V}$ whose restrictions to the fibres are regular. Denote by O the subsheaf of $\mathfrak{D}$ composed of germs of differentiable functions on $\mathscr{V}$ which are constant along the fibres of $\mathscr{V}$ (O is the sheaf of regular functions for the foliate structure of $\mathscr{V}$ whose "leaves" are the fibres). We note that O is the sheaf induced over $\mathscr{V}$ by the map $\varpi : \mathscr{V} \to B$ from the sheaf of germs of differentiable functions on B.

Let Θ, Ψ respectively be the sheaves of germs of differentiable sections of $\mathfrak{F}$, $\mathfrak{E}$ whose coefficients b_k^β in (2.2) are sections of $\mathfrak{D}$ (i.e. are regular along the fibres of $\mathscr{V}$). We thus have the exact sequence

$$0 \to \Theta \to \Psi \xrightarrow{j} \Lambda \to 0 \tag{3.1}$$

where Λ denotes the quotient sheaf. Now Ψ is a sheaf of germs of vector fields tangent to $\mathscr{V}$ and hence Ψ operates by differentiation on $\mathfrak{D}$. Let Π be the largest subsheaf of Ψ for which O is stable under these operations of Π. Clearly Θ is a subsheaf of Π, namely the subsheaf which annihilates O, and we obtain the exact sequence of sheaves

$$0 \to \Theta \to \Pi \xrightarrow{j} \mathrm{T} \to 0 \tag{3.2}$$

where T is the quotient sheaf. We call (3.2) the *fundamental sequence of sheaves* for the differentiable family of foliate structures.

Let T_B denote the sheaf of germs of differentiable sections of the tangent bundle of the manifold B, and let $\tilde{\mathrm{T}}$ be the sheaf induced over $\mathscr{V}$ from T_B by the map $\varpi : \mathscr{V} \to B$. There is a natural injection $\tilde{\mathrm{T}} \to \Lambda$ and the image of $\tilde{\mathrm{T}}$ in Λ is T. Thus $\Pi = j^{-1}(\mathrm{T})$ is the inverse image of T in Ψ under the map j of (3.1). For an arbitrary neighborhood U on B we observe that we have an isomorphism

$$H^0(\varpi^{-1}(U), \mathrm{T}) \cong H^0(U, \mathrm{T}_B). \tag{3.3}$$

Next, consider the fibre $V_t = \varpi^{-1}(t)$ of $\mathscr{V}$ over the point $t \in B$, and let Γ_t be the subsheaf of $\Theta|V_t$ (restriction of Θ to V_t) composed of those germs which vanish on V_t. We then have the exact sequence

$$0 \to \Gamma_t \to \Theta|V_t \to \Theta_t \to 0 \tag{3.4}$$

where Θ_t denotes the quotient sheaf. We denote by r_t the map defined by the commutative triangle

(3.5)

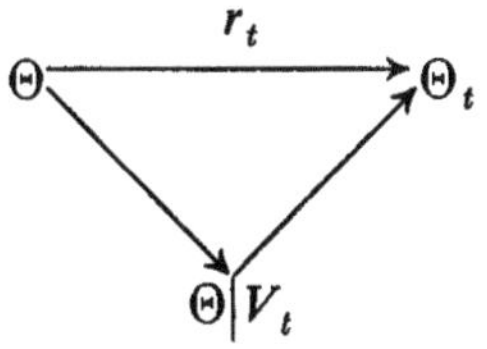

We call r_t the *restriction map* to the fibre V_t. The map r_t induces a restriction map r_t^* in the cohomology, namely

$$r_t^* : H^*(\mathscr{V}, \Theta) \to H^*(V_t, \Theta_t)$$

where, as usual, $H^*(\ldots)$ denotes the direct sum $\sum_s H^s(\ldots)$. In precisely the same way we define restriction maps for the other sheaves:

$$r_t : \Pi \to \Pi_t,\ r_t : \mathrm{T} \to \mathrm{T}_t, \text{ etc.}$$

We now make some remarks concerning the "exponentiation" of vector fields which will be applied in the next section. Given a differentiable vector field $\psi : \mathfrak{p} \to \psi(\mathfrak{p})$ on a differentiable manifold $\mathscr{U}$ and a subdomain $\mathscr{U}_1$ of $\mathscr{U}$ whose closure $\bar{\mathscr{U}}_1$ is compact and contained in $\mathscr{U}$, ψ determines a unique one-parameter family $\{e(s) | -\varepsilon < s < \varepsilon\}$ of differentiable homeomorphisms $e(s)$ of $\mathscr{U}_1$ into $\mathscr{U}$. In fact, in terms of the local coordinates $(x^1, \ldots, x^\nu, \ldots)$ on $\mathscr{U}$, $e(s) : x \to e(x, s)$ is determined from the simultaneous differential equations

$$\frac{d}{ds} e^\nu(x, s) = \psi^\nu(e(x, s)), \qquad e^\nu(x, 0) = x^\nu.$$

For simplicity we denote $e(s)$ by $\exp(s\psi)$.

Now let $\mathscr{V} \to B$ be a differentiable family of foliate structures and apply the above considerations to $\pi \in H^0(\mathscr{U}, \Pi)$ where $\mathscr{U}$ is a convex coordinate neighborhood on $\mathscr{V}$. In terms of the local coordinates $(x^\alpha, y^\beta, t^\gamma)$, π may be written in the form

$$\pi(x, y, t) = (a^1(x, y, t), \ldots, a^p(x, y, t); b^1(y, t), \ldots, b^q(y, t); c^1(t), \ldots, c^m(t)),$$

and

$$e(x, y, t, s) = (\ldots, f^\alpha(x, y, t, s), \ldots; \ldots, g^\beta(y, t, s), \ldots; \ldots h^\gamma(t, s), \ldots)$$

is determined by

$$\begin{cases} \dfrac{d}{ds} f^\alpha(x, y, t, s) = a^\alpha(f, g, h) \\ \dfrac{d}{ds} g^\beta(y, t, s) = b^\beta(g, h) \\ \dfrac{d}{ds} h^\gamma(t, s) = c^\gamma(h) \end{cases}$$

with the initial conditions

$$f^\alpha(x, y, t, 0) = x^\alpha, \quad g^\beta(y, t, 0) = y^\beta, \quad h^\gamma(t, 0) = t^\gamma.$$

It follows that

$$h(s) : t \to h(t, s)$$

defines a differentiable homeomorphism of $\varpi(\mathscr{U}_1)$ into $\varpi(\mathscr{U})$ and we have $h(s) = \exp(sc)$ where $c \in H^0(\varpi(\mathscr{U}), \mathrm{T}_B)$ is the canonical image of π under the homomorphism $j^* : H^0(\mathscr{U}, \Pi) \to H^0(\mathscr{U}, \mathrm{T}) \cong H^0(\varpi(\mathscr{U}), \mathrm{T}_B)$. Finally we observe that $g^\beta(y, t, s)$ is independent of the coordinates x^α and it is therefore regular along the fibres. Thus we obtain the following proposition:

PROPOSITION 3.1. *If $\pi \in H^0(\mathscr{U}, \Pi)$, then $e(s) = \exp(s\pi)$ is a differentiable homeomorphism of $\mathscr{U}_1$ into $\mathscr{U}$ which maps $V_t \cap \mathscr{U}_1$ biregularly into $V_{h(t,s)} \cap \mathscr{U}$ where $h(t, s) = \exp(sc)(t)$, $c = j^*(\pi)$.*

We conclude this section by making some comparisons with the case of complex analytic structure, and we remark first that the above sheaves Θ, Ψ are entirely analogous to the corresponding sheaves defined in [6] for complex analytic structure. Namely, in the complex analytic case, Θ, Ψ are respectively the sheaves of germs of differentiable sections of $\mathfrak{F}$, $\mathfrak{E}$ whose restrictions to each fibre of $\mathscr{V}$ are holomorphic. The definitions of Π, T, T_B are then the same as above. Finally, Proposition 3.1 is also entirely analogous to the corresponding proposition for complex structure (see [6]).

4. Local structure of a differentiable family

Let $\mathscr{V} \to B$ be a differentiable family of foliate structures and let Γ be an arbitrary sheaf over $\mathscr{V}$. It is convenient to introduce the corresponding sheaf $\mathscr{H}^*(\Gamma)$ of cohomology over B, namely the sheaf over B induced by the presheaf which assigns to each open set U of B the cohomology $H^*(\varpi^{-1}(U), \Gamma)$.

From the exact sequence (3.2) we obtain the following exact sequence of cohomology sheaves over B:

(4.1) $$0 \to \mathscr{H}^0(\Theta) \to \mathscr{H}^0(\Pi) \to \mathscr{H}^0(\mathrm{T}) \xrightarrow{\delta^*} \mathscr{H}^1(\Theta) \to \dots.$$

Since $\mathscr{H}^0(\mathrm{T}) \cong \mathrm{T}_B$ by (3.3), the map δ^* of (4.1) induces a map

$$\rho : \mathrm{T}_B \to \mathscr{H}^1(\Theta). \tag{4.2}$$

This map measures the magnitude of the dependence of the structure of the fibre V_t on the point $t \in B$, as we shall now show.

DEFINITION 4.1. *A differentiable family $\mathscr{V} \to B$ of $\mathscr{M}$-structures is said to be locally trivial if and only if every point $t_0 \in B$ has a neighborhood U together with a differentiable map $h : \varpi^{-1}(U) \to V_{t_0}$ whose restriction to each fibre V_t of $\varpi^{-1}(U)$ is a biregular map of V_t onto V_{t_0} in the sense of $\mathscr{M}$.*

We have the following theorem:

THEOREM 4.1. *A differentiable family of compact foliate manifolds is locally trivial if and only if the map ρ vanishes.*

This theorem, with the word "foliate" replaced by "complex analytic", is proved in [6] and the proof for foliate structures is the same if the exponentiation in [6] is replaced by that described at the end of the preceding section.

Local product structure, which lies over foliate structure, is defined by means of the category $\mathscr{M}$ whose objects are subdomains of $\mathbf{R}^n$ and whose maps are the differentiable maps with jacobians of the form

$$\begin{pmatrix} A & 0 \\ 0 & C \end{pmatrix}$$

where A is a $(p \times p)$-matrix, C a $(q \times q)$-matrix, $p + q = n$. Given a family $\mathscr{V} \to B$ of local product structures, the bundle $\mathfrak{F}$ of Section 2 splits: $\mathfrak{F} = \mathfrak{F}_x \oplus \mathfrak{F}_y$ into the sum of sub-bundles of vectors $\Sigma a_k^\alpha(\partial/\partial x_k^\alpha)$, $\Sigma b_k^\beta(\partial/\partial y_k^\beta)$ respectively. The fundamental sequence (2.5) thus becomes

$$0 \to \mathfrak{F}_x \oplus \mathfrak{F}_y \to \mathfrak{E} \to \mathfrak{E}/(\mathfrak{F}_x \oplus \mathfrak{F}_y) \to 0. \tag{4.3}$$

The sheaf Θ also splits: $\Theta = \Theta_x \oplus \Theta_y$ where Θ_x, Θ_y are respectively the sheaves of differentiable sections of $\mathfrak{F}_x$, $\mathfrak{F}_y$ whose components depend respectively on the x-coordinates, y-coordinates only. Theorem 4.1 remains valid for local product structure and examples show that the map ρ does not necessarily vanish (see [8], in particular the examples of local product structures on the 2-torus).

By restriction to a fibre V_t of $\mathscr{V} \to B$ we obtain a map

$$\rho_t : (\mathrm{T}_B)_t \to H^1(V_t, \Theta_t) \tag{4.2$_t$}$$

where $(\mathrm{T}_B)_t$ is the tangent space of the manifold B at the point t. In the case of complex analytic structure, the following two theorems can be proved for compact complex manifolds V_t by means of the theory of harmonic forms (see [6]):

(1) dim $H^1(V_t, \Theta_t)$ is finite and is an upper semi-continuous function of t.

(2) If dim $H^1(V_t, \Theta_t)$ is independent of t and if $\rho_t = 0$ for all t, then $\rho = 0$.

It follows in particular from (2) that dim $H^1(V_t, \Theta_t) = 0$ for all t implies $\rho = 0$ (theorem of Frölicher-Nijenhuis [4]).

For a family $\mathscr{V} \to B$ of compact local-product manifolds V_t, the examples of local product structures on the torus, cited above, show that dim $H^1(V_t, \Theta_t)$ may depend on t in a highly discontinuous way: dim $H^1(V_t, \Theta_t) = 2$ for almost all t, dim $H^1(V_t, \Theta_t) = \infty$ for an everywhere dense set of t. Hence (1) is false. With regard to (2) the situation is obscure. These difficulties are a reflection of the fact that the theory of harmonic forms for local product structure does not parallel that for complex analytic structure, and they emphasize the essential rôle of harmonic forms in representing cohomology classes.

5. Structures which leave a differential form invariant

The following are examples of primitive transitive real infinite continuous pseudo-groups, namely:

(1) The pseudo-group of all differentiable transformations in n variables $x^1, \ldots, x^n$ with non-vanishing jacobians.

(2) The pseudo-group of all differentiable transformations in n variables with jacobians equal to 1.

(3) The pseudo-group of all differentiable transformations in $n = 2m$ variables $x^1, \ldots, x^{2m} (m \geq 2)$ which leave invariant the exterior differential form $dx^1 \wedge dx^2 + dx^3 \wedge dx^4 + \ldots + dx^{2m-1} \wedge dx^{2m}$.

(4) The pseudo-group of all differentiable transformations in $n = 2m + 1$ variables $x^0, x^1, \ldots, x^{2m}$ which leave invariant, up to a non-vanishing factor, the Pfaffian form $dx^0 + x^1dx^2 - x^2dx^1 + \ldots + x^{2m-1}dx^{2m} - x^{2m}dx^{2m-1}$ (contact transformations).

(5) The pseudo-group of all differentiable transformations in n variables with constant jacobians, i.e. the pseudo-group which leaves invariant, up to a constant factor, the differential form $dx^1 \wedge dx^2 \wedge \ldots \wedge dx^n$.

(6) The pseudo-group of all differentiable transformations in $n = 2m$ variables $x^1, \ldots, x^{2m} (m \geq 2)$ which leave invariant, up to a constant factor, the exterior differential form $dx^1 \wedge dx^2 + dx^3 \wedge dx^4 + \ldots + dx^{2m-1} \wedge dx^{2m}$.

Of these six types only the first four are *simple*, i.e. contain no proper normal pseudo sub-group different from the identity. In the complex analytic case, the above six infinite continuous pseudo-groups constitute all types of primitive transitive infinite continuous pseudo-groups (E. Cartan [1]).

We denote the coordinates of real n-space $\mathbf{R}^n$ by $x^1, x^2, \ldots, x^n$, and consider on $\mathbf{R}^n$ a differential form ψ of degree p, namely

(5.1) $$\psi = \psi(x) = \frac{1}{p!} \Sigma\, \psi_{ij\ \ k}(x) dx^i \wedge dx^j \wedge \ldots \wedge dx^k.$$

Let $\mathscr{M}(\psi)$ be the category whose objects are the subdomains of $\mathbf{R}^n$ and whose maps are all differentiable maps leaving the form ψ invariant. (An $\mathscr{M}(\psi)$-manifold is one which possesses an $\mathscr{M}(\psi)$-structure in the sense of Definition 1.1.) Of the above simple pseudo-groups, (2) leaves invariant the form $dx^1 \wedge \ldots \wedge dx^n$ while (3) reproduces the form $dx^1 \wedge dx^2 + dx^3 \wedge dx^4 + \ldots + dx^{2m-1} \wedge dx^{2m}$ of $\mathbf{R}^{2m}$.

Now let $f_t : y \rightarrow x = f(y, t)$ be a differentiable map depending differentiably on t and, for simplicity, assume that $-\infty < t < +\infty$. In terms of the coordinates we write $x^\alpha = f^\alpha(y, t)$ $(\alpha = 1, 2, \ldots, n)$. For fixed y and varying t, $x = f(y, t)$ describes an arc on x-space ($\mathbf{R}^n$ with coordinates $x^1, \ldots, x^n$) with the tangent $\theta^n = (\theta^1, \ldots, \theta^n)$ where

$$\theta^i(x, t) = \frac{\partial}{\partial t} f^i(y, t), \qquad x = f(y, t).$$

As is well known, θ operates by differentiation on the differential forms of x-space and defines a derivation in the exterior algebra of these differential forms, i.e. $\theta \cdot (\varphi \wedge \psi) = (\theta \cdot \varphi) \wedge \psi + \varphi \wedge (\theta \cdot \psi)$. Denoting by f_t^* the homorphism of the exterior algebra of differential forms which is induced by the map f_t, we have the formula

(5.2) $$\frac{\partial}{\partial t} f_t^* \psi = f_t^*(\theta \cdot \psi).$$

In fact, the operation of contraction with the vector field θ defines an anti-deviation $i(\theta)$ of degree -1 of the exterior algebra of the differential forms on x-space, where $i(\theta)$ vanishes on forms of degree 0 and is defined on forms of degree $p > 0$ by

$$i(\theta)\psi = \frac{1}{(p-1)!} \sum_{i,j,\ \ ,k} \theta^i \psi_{ij\ \ k} dx^j \wedge \ldots \wedge dx^k.$$

We have the classical formula

(5.3) $$\theta \cdot \psi = d(i(\theta)\psi) + i(\theta) d\psi$$

where d denotes exterior differentiation. We reduce formula (5.2) to (5.3) as follows. By definition,

$$f_t^* \psi = \frac{1}{p!} \sum \psi_{ij\ \ k}(f(y, t)) df^i \wedge df^j \wedge \ldots \wedge df^k.$$

Since

$$df^j(y, t) = \sum_h \frac{\partial f^j(y, t)}{\partial y^h} dy^h,$$

we have

$$d\theta^j = \frac{\partial}{\partial t} df^j(y, t) = \sum_h \frac{\partial f^j(y, t)}{\partial y^h \partial t} dy^h$$

and hence

$$\frac{\partial}{\partial t} f_t^* \psi = \frac{1}{p!} \sum \frac{\partial}{\partial t} \psi_{hj\quad k}(f(y, t)) dx^h \wedge dx^j \wedge \ldots \wedge dx^k$$

$$+ \frac{1}{(p-1)!} \sum \psi_{ij\quad k}(x) d\theta^i \wedge dx^j \wedge \ldots \wedge dx^k.$$

On the other hand,

$$\frac{1}{(p-1)!} \sum \psi_{ij\quad k} d\theta^i \wedge dx^j \wedge \ldots \wedge dx^k$$

$$= \frac{1}{(p-1)!} \sum d(\theta^i \psi_{ij\quad k} dx^j \wedge \ldots \wedge dx^k)$$

$$- \frac{1}{(p-1)!} \sum \theta^i d(\psi_{ij\quad k} dx^j \wedge \ldots \wedge dx^k)$$

$$= \frac{1}{(p-1)!} \sum d(\theta^i \psi_{ij\quad k} dx^j \wedge \ldots \wedge dx^k)$$

$$- \frac{1}{p!} \sum \theta^i \left(\frac{\partial}{\partial x^h} \psi_{ij\quad k} - \frac{\partial}{\partial x^j} \psi_{ih\quad k} + \ldots \right) dx^h \wedge dx^j \wedge \ldots \wedge dx^k.$$

Therefore

$$\frac{\partial}{\partial t} f_t^* \psi = d \left\{ \frac{1}{(p-1)!} \sum \theta^i \psi_{ij\quad k} dx^j \wedge \ldots \wedge dx^k \right\}$$

$$+ \frac{1}{p!} \sum \theta^i \left(\frac{\partial}{\partial x^i} \psi_{hj\quad k} - \frac{\partial}{\partial x^h} \psi_{ij\quad k} + \ldots \right) dx^h \wedge dx^j \wedge \ldots \wedge dx^k$$

$$= f_t^*(\theta \cdot \psi)$$

by (5.3), q.e.d.

EXAMPLES. (i) Consider the n-form

$$\psi = dx^1 \wedge dx^2 \wedge \ldots \wedge dx^n. \tag{5.4}$$

We have $f_t^* \psi = J(y, t) dy^1 \wedge \ldots \wedge dy^n$ where

$$J(y, t) = \det\left(\frac{\partial x^i}{\partial y^j} \right) = \begin{vmatrix} \dfrac{\partial x^1}{\partial y^1} & \cdots & \dfrac{\partial x^n}{\partial y^1} \\ \cdots & \cdots & \cdots \\ \dfrac{\partial x^1}{\partial y^n} & \cdots & \dfrac{\partial x^n}{\partial y^n} \end{vmatrix}.$$

Hence

$$\frac{\partial}{\partial t}(f_t^* \psi) = \frac{\partial J}{\partial t} dy^1 \wedge \ldots \wedge dy^n.$$

On the other hand

$$i(\theta)\psi = \sum_k (-1)^{k-1}\theta^k dx^1 \wedge \ldots \wedge dx^{k-1} \wedge dx^{k+1} \wedge \ldots \wedge dx^n$$

and therefore

$$d(i(\theta)\psi) = (\nabla \cdot \theta) dx^1 \wedge \ldots \wedge dx^n$$

where

$$\nabla \cdot \theta = \sum_{k=1}^{n} \frac{\partial \theta^k}{\partial x^k}$$

is the "divergence" of the vector field θ. Since $f_t^*\psi = J dy^1 \wedge \ldots \wedge dy^n$, we obtain from (5.2) and (5.3) the formula:

$$\frac{\partial J}{\partial t} = J \cdot (\nabla \cdot \theta) \tag{5.5}$$

or, if $J \neq 0$, $\partial \log J/\partial t = \nabla \cdot \theta$. Thus J is independent of t (case of the pseudo-group (2)) if and only if

$$\nabla \cdot \theta = 0. \tag{5.6}$$

More generally (case of the pseudo-group (5)), the equation

$$\nabla \cdot \theta = \text{function of } t \tag{5.7}$$

is equivalent to the condition that $J(y, t)$ is independent of y.

The following formula is easily verified:

$$\frac{\partial J}{\partial y^k} = J \cdot \sum_{j=1}^{n} \frac{\partial}{\partial x^j}\left(\frac{\partial x^j}{\partial x^k}\right) \qquad (k = 1, \ldots, n). \tag{5.8}$$

Hence J is constant (i.e. independent of y) if and only if

$$\sum_{j=1}^{n} \frac{\partial}{\partial x^j}\left(\frac{\partial x^j}{\partial x^k}\right) = 0. \tag{5.9}$$

It follows immediately from (5.9) that, if $y \to x = f(y, t)$ is, for each fixed t, a transformation of the pseudo-group (5), then

$$\sum_{j=1}^{n} \frac{\partial \theta^j}{\partial x^j} = \sum_{k=1}^{n} \frac{\partial \varphi^k}{\partial y^k}$$

where the φ^k are the components of the vector field θ expressed in terms of the coordinates $y^1, \ldots, y^n$, i.e.

$$\theta^j = \sum_k \frac{\partial x^j}{\partial y^k} \varphi^k.$$

Thus the expression $\nabla \cdot \theta$ is invariant under transformations of the pseudo-group (5).

(ii) Consider on $\mathbf{R}^{2m}$ the 2-form

$$\psi = dx^1 \wedge dx^2 + dx^3 \wedge dx^4 + \dots + dx^{2m-1} \wedge dx^{2m} = \tfrac{1}{2} \sum_{j,k} \varepsilon_{jk} dx^j \wedge dx^k \tag{5.10}$$

where

$$\varepsilon_{jk} = \begin{cases} 1, & j = 2i-1, \quad k = 2i, \\ -1, & j = 2i, \quad k = 2i-1, \\ 0, & \text{otherwise.} \end{cases}$$

We have

$$i(\theta)\psi = \sum_{j,k} \theta^j \varepsilon_{jk} dx^k = \theta^1 dx^2 - \theta^2 dx^1 + \theta^3 dx^4 - \theta^4 dx^3 + \dots,$$

and hence

$$d(i(\theta)\psi) = d\theta^1 \wedge dx^2 + dx^1 \wedge d\theta^2 + d\theta^3 \wedge dx^4 + dx^3 \wedge d\theta^4 + \dots .$$

Thus, since $d\psi = 0$, we obtain from (5.2) and (5.3) the formula

$$\frac{\partial}{\partial t}(f_t^* \psi) = d(i(\theta)\psi) = d\theta^1 \wedge dx^2 + dx^1 \wedge d\theta^2 + \dots . \tag{5.11}$$

We conclude that the transformation $y \to x = f(y, t)$ leaves the form (5.10) invariant if and only if

$$\sum_k (d\theta^{2k-1} \wedge dx^{2k} + dx^{2k-1} \wedge d\theta^{2k}) = 0 \tag{5.12}$$

(a system of equations).

Now let $\mathscr{V} \to B$ be a differentiable family of *compact* $\mathscr{M}(\psi)$-manifolds where ψ is a differential form of degree p, that is to say, there is a covering $\{\mathscr{U}_i\}$ of $\mathscr{V}$ by domains $\mathscr{U}_i$ with coordinates $(x_i^1, \dots, x_i^n, t)$ where t now denotes a point of the differentiable parameter manifold B, such that the transformations $x_i = h_{ik}(x_k, t)$ leave invariant the form ψ. We recall that $\mathscr{V}$ and B are always assumed to be orientable (see the remark following Definition 2.1). Let Π be the sheaf of germs of vector fields tangent to $\mathscr{V}$ of the form $\pi_i(x, t) = (\theta_i^1(x, t), \dots, \theta_i^n(x, t), v(t))$ where the θ_i^j are the components of a vector field θ tangent to the fibres of $\mathscr{V}$ which satisfies the condition

$$\theta \cdot \psi = 0 \tag{5.13}$$

and where $v(t)$ is a "horizontal" vector field tangent to B. Let Θ be the subsheaf of Π composed of germs of vector fields $(\theta^1, \dots, \theta^n, 0)$. As in Section 3 we have the exact sequence of sheaves

$$0 \to \Theta \to \Pi \to T \to 0$$

and the map $\rho : \mathrm{T}_B \to \mathscr{H}^1(\Theta)$. It is readily verified that the analogue of Proposition 3.1 is valid for differentiable families of $\mathscr{M}(\psi)$-manifolds and hence we obtain

THEOREM 5.1. *A differentiable family of compact $\mathscr{M}(\psi)$-manifolds is locally trivial if and only if the map ρ vanishes.*

Similarly, let $\mathscr{M}_c(\psi)$, $\mathscr{M}_g(\psi)$ be the categories whose objects are the subdomains of $\mathbf{R}^n$ and whose maps are all differentiable maps leaving the form ψ invariant up to a constant factor and up to a variable (non-vanishing) factor respectively. Theorem 5.1 remains valid with $\mathscr{M}_c(\psi)$ and $\mathscr{M}_g(\psi)$ replacing $\mathscr{M}(\psi)$; the only difference is that equation (5.13) is to be replaced by $\theta \cdot \psi = c\psi$, $\theta \cdot \psi = g\psi$ where $c = c(t)$ is a function of t and where $g = g(x, t)$ is a function of both x and t. Thus, in particular, Theorem 5.1 is valid for all the structures defined by the primitive infinite pseudo-groups (1)–(6).

Next, we consider in greater detail families of compact $\mathscr{M}(\psi)$-structures where ψ is the n-form (5.4). Let, then, $\mathscr{V} \to B$ be a differentiable family of compact $\mathscr{M}(\psi)$-manifolds where ψ is the form (5.4). The integral

$$v(t) = \int_{V_t} dx^1 \wedge \ldots \wedge dx^n \tag{5.14}$$

over the fibre V_t then has an invariant meaning and it will be called the volume of V_t. Denote by O_B the sheaf of germs of differentiable functions on B and, for any coordinate domain $U \subset B$, let $T_B(U) = H^0(U, T_B)$ be the module of differentiable vector fields over U. Finally, let Θ be the above-defined sheaf of germs of vector fields along the fibres of $\mathscr{V}$; in the case of the particular structure under consideration the equation (5.13) reduces to (5.6), i.e. Θ is the sheaf of vector fields θ satisfying $\nabla \cdot \theta = 0$.

THEOREM 5.2. *We have the isomorphism $\mathscr{H}^1(\Theta) \cong O_B$ and, for $u \in \mathrm{T}_B(U)$, $\rho(u) \in H^1(\mathscr{V} \mid U, \Theta) \cong H^0(U, O_B)$ may be identified with the function*

$$u \cdot v : t \to \sum_\nu u^\nu(t) \frac{\partial v(t)}{\partial t^\nu}$$

where $u^1, \ldots, u^m$ are the components of u referred to the coordinates $t^1, \ldots, t^m$ covering U, and v is defined by (5.14).

Thus, in particular, the family $\mathscr{V} \to B$ is locally trivial if and only if all the fibres V_t of $\mathscr{V}$ have the same volume.

PROOF. Denote by Ξ the sheaf over $\mathscr{V}$ of germs of vector fields ξ along the fibres of $\mathscr{V}$, and let Φ^p be the sheaf over $\mathscr{V}$ of differential p-forms along the fibres of $\mathscr{V}$, i.e. Φ^p is the sheaf of germs of differentiable sections of the bundle $\bigwedge_p \mathfrak{F}^*$ where $\mathfrak{F}^*$ is the dual of the bundle $\mathfrak{F}$ along the fibres of $\mathscr{V}$ and where $\bigwedge_p$ denotes the p-tuple exterior product of the vector bundle $\mathfrak{F}^*$.

Since the form (5.4) is invariant along the fibres of $\mathscr{V}$ (by hypothesis), we have a map $\Xi \to \Phi^{n-1}$ defined by contraction of Ξ with $\psi = dx^1 \wedge \ldots \wedge dx^n$, namely: $\xi = (\xi^1, \ldots, \xi^k, \ldots, \xi^n) \in \Xi$ goes into the $(n-1)$-form $\varphi(\xi) = i(\xi)\psi = \Sigma(-1)^{k-1}\xi^k dx^1 \wedge \ldots \wedge dx^{k-1} \wedge dx^{k+1} \wedge \ldots \wedge dx^n$. We remark that, under a transformation of coordinates, $\varphi(\xi)$ is multiplied by the jacobian which, however, is equal to 1. This map is clearly an isomorphism and we have the commutative diagram

$$\begin{array}{ccc} \Xi & \cong & \Phi^{n-1} \\ \nabla\downarrow & & d\downarrow \\ \Phi^0 & \cong & \Phi^n \end{array}$$

Thus we obtain the exact commutative diagram

$$\begin{array}{ccccccccc} 0 & \to & \Theta & \longrightarrow & \Xi & \xrightarrow{\nabla} & \Phi^0 & \to & 0 \\ & & \| & & \| & & \| & & \\ 0 & \to & \Psi^{n-1} & \to & \Phi^{n-1} & \xrightarrow{d} & \Phi^n & \to & 0 \end{array} \tag{5.15}$$

where Ψ^{n-1} denotes the subsheaf of Φ^{n-1} composed of germs of d-closed $(n-1)$-forms and where the vertical double lines indicate isomorphisms. From the exact cohomology sequence corresponding to the second row of (5.15) we infer that, for any open set $U \subset B$,

$$H^1(\mathscr{V} \mid U, \Theta) \cong H^1(\mathscr{V} \mid U, \Psi^{n-1}) \cong H^0(\mathscr{V} \mid U, \Phi^n)/dH^0(\mathscr{V} \mid U, \Phi^{n-1}).$$

We plainly have

$$H^0(V_t, \Phi_t^n)/dH^0(V_t, \Phi_t^{n-1}) \cong \mathbf{C} \qquad (t \in B)$$

where V_t is the fibre of $\mathscr{V}$ over the point $t \in B$ and where Φ_t^p denotes the restriction (in the sense of Section 3) of the sheaf Φ^p to the fibre V_t, i.e. $\Phi_t^p = r_t(\Phi^p)$ (compare (3.5)). Hence we can apply the real analogue of the theory of harmonic forms along the fibres of $\mathscr{V}$ which was developed in Section 2 of the paper [6] to prove that $\mathscr{H}^1(\Theta) \cong O_B$.

To obtain the second statement of Theorem 5.2, we let $\{\mathscr{U}_i\}$ be a covering of $\mathscr{V} \mid U$ by domains $\mathscr{U}_i$ with coordinates $(x_i^1, \ldots, x_i^n, t)$ where $x_i^\alpha = h_{ik}^\alpha(x_k, t)$ $(\alpha = 1, 2, \ldots, n)$. Now $\mathscr{V}$ is differentiably locally trivial by hypothesis, i.e. there exist differentiable functions such that $f_i^\alpha(y_i, t) = h_{ik}^\alpha(f_k(y_k, t), t)$, $y_i^\alpha = h_{ik}^\alpha(y_k, 0)$. For simplicity we suppose henceforth that $\dim B = 1$ and that the vector field $u \in T_B(U)$ is simply $\partial/\partial t$ where t is the parameter of B. Then $\theta = \rho(u) \in H^1(\mathscr{V} \mid U, \Theta)$ is represented on the nerve of the covering by the 1-cocycle $\{\theta_{ik}\}$ where θ_{ik} is the section of the sheaf Θ over $\mathscr{U}_i \cap \mathscr{U}_k$ whose components θ_{ik}^α referred to the coordinates $x_i^1, \ldots, x_i^n$ are defined by

$$\theta_{ik}^\alpha(x_i, t) = \frac{\partial}{\partial t} h_{ik}^\alpha(x_k, t).$$

If we let ξ be the section of the sheaf Ξ over $\mathscr{U}_\iota$ whose components with respect to the coordinates $x_\iota^1, \ldots, x_\iota^n$ are

$$\xi_\iota^\alpha = \frac{\partial}{\partial t} f_\iota^\alpha(y_\iota, t),$$

then

$$\theta_{\iota k} = \xi_k - \xi_\iota \quad \text{in} \quad \mathscr{U}_\iota \cap \mathscr{U}_k.$$

The exact cohomology sequence corresponding to the first row of the diagram (5.15) is

$$\ldots \to H^0(\mathscr{V} \mid U, \Xi) \to H^0(\mathscr{V} \mid U, \Phi^0) \xrightarrow{\delta^*} H^1(\mathscr{V} \mid U, \Theta) \to 0 \to \ldots.$$

Here we have used the fact that $H^1(\mathscr{V} \mid U, \Xi)$ vanishes because Ξ is a *fine* sheaf. Now $\theta = \delta^* \varphi$ where $\varphi \in H^0(\mathscr{V} \mid U, \Phi^0)$ is the function which coincides in $\mathscr{U}_\iota$ with $\nabla \cdot \xi_\iota$. Since $df_\iota^1 \wedge df_\iota^2 \wedge \ldots \wedge df_\iota^n = J_\iota(y, t) dy_\iota^1 \wedge \ldots \wedge dy_\iota^n$ where J_ι denotes the jacobian, we conclude that $J_\iota(y, t) = J_k(y, t)$ in $\mathscr{U}_\iota \cap \mathscr{U}_k$ and, by (5.5),

$$\frac{1}{J_\iota} \frac{\partial}{\partial t} J_\iota = \nabla \cdot \xi_\iota = \varphi.$$

The homomorphism

$$H^0(\mathscr{V} \mid U, \Phi^0) \to H^0(\mathscr{V} \mid U, O_B) \cong H^0(\mathscr{V} \mid U, \Phi^n)/dH^0(\mathscr{V} \mid U, \Phi^{n-1})$$

is simply given by

$$\varphi \to \int_{V_t} \varphi dx_\iota^1 \wedge \ldots \wedge dx_\iota^n$$

where

$$\begin{aligned}
\int_{V_t} \varphi dx_\iota^1 \wedge \ldots \wedge dx_\iota^n &= \int_{V_t} \frac{1}{J_\iota} \frac{\partial}{\partial t} J_\iota \cdot dx_\iota^1 \wedge \ldots \wedge dx_\iota^n \\
&= \int_{V_t} \frac{\partial}{\partial t} J_\iota \cdot dy_\iota^1 \wedge \ldots \wedge dy_\iota^n \\
&= \frac{\partial}{\partial t} \int_{V_t} J_\iota(y, t) dy_\iota^1 \wedge \ldots \wedge dy_\iota^n \\
&= \frac{\partial}{\partial t} \int_{V_t} dx_\iota^1 \wedge \ldots \wedge dx_\iota^n = \frac{\partial}{\partial t} v(t),
\end{aligned}$$

q.e.d.

Finally, let $\mathscr{V} \to B$ be a differentiable family of compact $\mathscr{M}_c(\psi)$-manifolds where ψ is the differential form (5.4), i.e. there is a covering $\{\mathscr{U}_\iota\}$ of $\mathscr{V}$ by domains $\mathscr{U}_\iota$ with coordinates $(x_\iota^1, \ldots, x_\iota^n, t)$, $t \in B$, where the jacobians $J_{\iota k} = J_{\iota k}(t) = \det(\partial x_\iota^\alpha / \partial x_k^\beta)$ depend only on the point t. Denote by Ξ the sheaf of germs of differentiable vector fields φ tangent to the fibres of $\mathscr{V}$, and let $\nabla \cdot \varphi = \Sigma_\alpha \partial \varphi_\iota^\alpha / \partial x_\iota^\alpha$ where $\varphi_\iota^1, \ldots, \varphi_\iota^n$ are the components of φ referred to the coordinates $x_\iota^1, \ldots, x_\iota^n$. Let Θ_c be the subsheaf of Ξ

composed of germs of vector fields φ whose divergence $\nabla \cdot \varphi$ is a differentiable function depending only on t and let Θ be the subsheaf of Θ_c composed of germs of vector fields φ whose divergence vanishes, i.e. $\nabla \cdot \varphi = 0$. As in Section 3 let O be the sheaf of germs of differentiable functions on $\mathscr{V}$ which are constant along the fibres of $\mathscr{V}$ and, as above, let Φ^p denote the sheaf over $\mathscr{V}$ of germs of differential forms along the fibres of $\mathscr{V}$ of degree p and let Ψ'^p be the subsheaf of germs of d-closed forms where d is exterior differentiation along the fibres. Then we have the exact commutative diagram:

$$\begin{array}{ccccccccc}
 & & 0 & & & & & & \\
 & & \uparrow & & & & & & \\
 & & O & & 0 & & 0 & & \\
 & & \uparrow & & \uparrow & & \uparrow & & \\
0 & \longrightarrow & \Theta_c & \longrightarrow & \Xi & \xrightarrow{d\cdot\nabla} & \Psi'^1 & \longrightarrow & 0 \\
 & & \uparrow & & \uparrow & & \uparrow{\scriptstyle d} & & \\
0 & \longrightarrow & \Theta & \longrightarrow & \Xi & \xrightarrow{\nabla} & \Phi^0 & \longrightarrow & 0 \\
 & & \uparrow & & \uparrow & & \uparrow & & \\
 & & 0 & & 0 & & O & & \\
 & & & & & & \uparrow & & \\
 & & & & & & 0 & &
\end{array} \tag{5.16}$$

The corresponding exact commutative diagram of cohomology sheaves over B is

$$\begin{array}{ccccccccc}
 & & & & \vdots & & & & \\
 & & & & \uparrow & & & & \\
\cdots \longrightarrow & \mathscr{H}^0(\Xi) & \xrightarrow{d\cdot\nabla} & & \mathscr{H}^0(\Psi'^1) & \longrightarrow & \mathscr{H}^1(\Theta_c) & \longrightarrow & 0 \\
 & \Vert & & & {\scriptstyle d}\uparrow & & & & \\
\cdots \longrightarrow & \mathscr{H}^0(\Xi) & \xrightarrow{\nabla} & & \mathscr{H}^0(\Phi^0) & \longrightarrow & \cdots & & \\
 & & & & \uparrow & & & & \\
 & & & & \vdots & & & &
\end{array} \tag{5.17}$$

where the vertical double lines denote the identity map.

If the 1-cocycle $\{\log J_{ik}(t)\}$ on V_t is homologous to zero, then V_t is an $\mathscr{M}(\psi)$-manifold, a case which we discussed above. We therefore suppose that $\{\log J_{ik}(t)\}$ is non-homologous to zero on each V_t. With the help of the theory of harmonic forms along the fibres of $\mathscr{V}$ (compare Section 2 of [6]) it can be shown that $\nabla.\mathscr{H}^0(\Xi) = \mathscr{H}^0(\Phi^0)$ and from this we obtain the following:

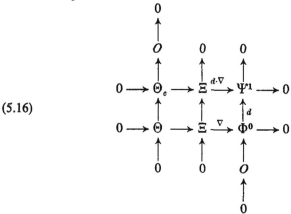

Theorem 5.3. *We have the isomorphism* $\mathscr{H}^1(\Theta_c) \cong \mathscr{H}^1(O)$ *and, for* $u \in T_B(U)$, $\rho(u) \in H^1(\mathscr{V} \mid U, \Theta_c) \cong H^1(\mathscr{V} \mid U, O)$ *may be identified with*

the cohomology class represented by the 1-cocycle $\{u \cdot \log J_{ik}(t)\}$ *with coefficients in* O *on the nerve of the covering* $\{\mathscr{U}_\iota\}$ *of* $\mathscr{V} \mid U$.

A proof of this theorem will be omitted.

6. Complex analytic structures which leave a differential form invariant

There are four types of simple transitive infinite continuous pseudo-groups of complex analytic transformations corresponding to the four types of simple infinite continuous pseudo-groups of real differentiable transformations which were described at the beginning of Section 5 (see also [3]); in particular, the pseudo-group of all complex analytic transformations in n complex variables $z^1, \ldots, z^n$ with jacobians equal to 1 is simple. We may introduce an arbitrary holomorphic differential form of type $(p, 0)$ on complex n-space $\mathbf{C}^n$, namely

$$\psi = \psi(z) = \frac{1}{p!}\, \Sigma \psi_{ij\ldots k}(z) dz^i \wedge dz^j \wedge \ldots \wedge dz^k, \tag{6.1}$$

and consider the category $\mathscr{M}(\psi)$ whose objects are the subdomains of $\mathbf{C}^n$ and whose maps are all complex analytic maps leaving ψ invariant. For simplicity, we shall confine our attention to the n-form

$$\psi = dz^1 \wedge dz^2 \wedge \ldots \wedge dz^n, \tag{6.2}$$

which corresponds to the simple pseudo-group mentioned above.

Let $\mathscr{V} \to B$ be a differentiable family of compact, complex $\mathscr{M}(\psi)$-structures where ψ is the form (6.2), and let $\mathfrak{F}$ be the bundle along the fibres of $\mathscr{V}$ which was defined for families of complex analytic structures at the end of Section 2. We denote by Ξ the sheaf of germs of differentiable sections of the bundle $\mathfrak{F}$ which are holomorphic along the fibres of $\mathscr{V}$ (Ξ is the fundamental sheaf which is denoted by Θ in the paper [6]), and we denote by Θ the subsheaf of Ξ which is composed of germs θ satisfying

$$\nabla \cdot \theta = \sum_{\alpha=1}^{n} \frac{\partial \theta^\alpha}{\partial z^\alpha} = 0 \tag{6.3}$$

(compare (5.6)). Finally, denote by $\mathfrak{D}$ the sheaf of germs of differentiable functions on $\mathscr{V}$ which are holomorphic along the fibres. We have the exact sequence

$$0 \longrightarrow \Theta \xrightarrow{\iota} \Xi \xrightarrow{\nabla} \mathfrak{D} \longrightarrow 0, \tag{6.4}$$

and hence we obtain the exact commutative diagram

$$\begin{array}{ccccccccc} \ldots \longrightarrow & \mathscr{H}^0(\Xi) & \longrightarrow & \mathscr{H}^0(\mathfrak{D}) & \xrightarrow{\delta^*} & \mathscr{H}^1(\Theta) & \xrightarrow{\iota^*} & \mathscr{H}^1(\Xi) & \longrightarrow \ldots \\ & & & & & \uparrow \rho & \nearrow \eta & & \\ & & & & & \mathrm{T}_B & & & \end{array} \tag{6.5}$$

We remark that there is no difficulty in extending Theorem 5.1 to differentiable families of compact, *complex* $\mathscr{M}(\psi)$-manifolds; in particular, the above family $\mathscr{V} \to B$ is locally trivial if and only if the map ρ vanishes.

We affix a subscript t to the sheaves Θ, Ξ to denote the restriction (in the sense of Section 3) to the fibre V_t of $\mathscr{V}$ over the point $t \in B$, and we write Ω_t for the corresponding restriction of $\mathfrak{O}$ to V_t (Ω_t is the sheaf of germs of holomorphic functions on V_t). Then we have the diagram obtained by restricting (6.5) to the fibre V_t, namely:

$$\begin{array}{l} \ldots \longrightarrow H^0(V_t, \Xi_t) \longrightarrow H^0(V_t, \Omega_t) \xrightarrow{\delta_t^*} H^1(V_t, \Theta_t) \xrightarrow{\iota_t^*} H^1(V_t, \Xi_t) \longrightarrow \ldots \\ (6.5)_t \qquad\qquad\qquad\qquad\qquad\qquad\qquad\quad \uparrow \rho_t \quad \nearrow \eta_t \\ \qquad\qquad\qquad\qquad\qquad\qquad\qquad\qquad\quad (\mathrm{T}_B)_t \end{array}$$

Now let $\mathscr{W} \to \mathscr{V}$ be a differentiable vector bundle over $\mathscr{V}$ whose restriction $W_t = \mathscr{W} \mid V_t$ to each fibre V_t of $\mathscr{V}$ is a holomorphic vector bundle, and denote by $\mathfrak{O}(\mathscr{W})$ the sheaf of germs of differentiable sections of $\mathscr{W}$ whose restrictions to the fibres of $\mathscr{V} \to B$ are *holomorphic*. We define the restriction $\Omega(W_t)$ of $\mathfrak{O}(\mathscr{W})$ to the fibre V_t in the same manner that we defined the restriction Θ_t of Θ to V_t in Section 3 (compare (3.5)); then $\Omega(W_t)$ is the sheaf over V_t of germs of holomorphic sections of the bundle W_t. The following lemma is part (ii) of the fundamental Theorem 2.2 of [6] (whose proof is based on the theory of harmonic forms along the fibres of $\mathscr{V}$):

LEMMA 6.1. *Suppose that* $H^q(V_t, \Omega(W_t))$ *is independent of* $t \in B$ *and let* U *be an arbitrary open set in* B. *Then* $r_t^*(\varphi) = 0$ *for all* $t \in U$ *implies* $\varphi = 0$ *where* $\varphi \in H^q(\mathscr{V} \mid U, \mathfrak{O}(\mathscr{W}))$:

Using this lemma, we now prove:

THEOREM 6.1. *Assume that* dim $H^1(V_t, \Xi_t)$ *is independent of* $t \in B$. *and that* $H^0(V_t, \Xi_t) = 0$, $\rho_t = 0$ *for each* $t \in B$. *Then* $\mathscr{V} \to B$ *is locally trivial.*

PROOF. Since $\rho_t = 0$ by hypothesis, we conclude that $\eta_t = \iota_t^* \circ \rho_t = 0$ for each $t \in B$. Since dim $H^1(V_t, \Xi_t)$ is assumed to be independent of t, we infer from Lemma 6.1 that $\eta = 0$. But $\eta = \iota^* \circ \rho$; therefore $\iota^*(\rho(\mathrm{T}_B)) = 0$. Since dim $H^0(V_t, \Xi_t) = 0$, Lemma 6.1 implies that $\mathscr{H}^0(\Xi) = 0$; therefore ρ can be lifted back uniquely to a map $\sigma : \mathrm{T}_B \to \mathscr{H}^0(\mathfrak{O})$. But $0 = \delta_t^* \circ \sigma_t$ where δ_t^* is injective; hence $\sigma_t = 0$. Since $H^0(V_t, \Omega_t) \cong \mathbf{C}$, i.e. dim $H^0(V_t, \Omega_t) = 1$, a final application of Lemma 6.1 shows that $\sigma = 0$ and therefore, since δ^* is injective, that $\rho = 0$, q.e.d.

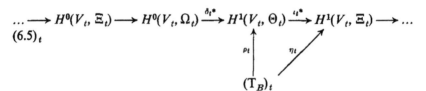

We remark that Theorem 6.1 results from the theory of harmonic forms applied to the cohomology with coefficients in the sheaf Ξ, and the question arises whether harmonic theory cannot be applied directly to the cohomology with coefficients in Θ. The answer is *negative*. In fact, the application of the harmonic theory makes use of the isomorphism of $H^q(V_t, \Xi_t)$ with the $\bar{\partial}_t$-cohomology of differential forms of type $(0, q)$ on V_t with values (coefficients) in the holomorphic tangent bundle of V_t (Dolbeault isomorphism). Here $\bar{\partial}_t$ denotes the differential operator of type (0, 1) on V_t (compare [6]). It is a trivial consequence of this isomorphism that $H^q(V_t, \Xi_t)$ vanishes whenever q exceeds the complex dimension of V_t. On the other hand, an example of an algebraic surface V_t exists for which $H^3(V_t, \Theta_t)$ does not vanish. In fact, dropping all subscripts t for simplicity, we consider a surface V of order 4 in $P_3(\mathbf{C})$ (complex projective 3-space), for example the surface defined in terms of the homogeneous coordinates z_0, z_1, z_2, z_3 of $P_3(\mathbf{C})$ by the equation $z_0^4 + z_1^4 + z_2^4 + z_3^4 = 0$. We have the exact sequence.

$$0 \to \Theta \to \Xi \xrightarrow{\nabla} \Omega \to 0$$

where Θ is the subsheaf of the sheaf Ξ of holomorphic vector fields which is composed of those germs having a vanishing divergence and where Ω is the sheaf of germs of holomorphic functions on V. The corresponding exact cohomology sequence is

$$\ldots \to H^2(V, \Xi) \to H^2(V, \Omega) \to H^3(V, \Theta) \to H^3(V, \Xi) \to \ldots$$

where, as may be verified, $H^2(V, \Xi) = H^3(V, \Xi) = 0$, $H^2(V, \Omega) \cong \mathbf{C}$. Thus $H^3(V, \Theta) \cong \mathbf{C}$, q.e.d.

References

[1] Cartan, E., Les groupes de transformations continus, infinis, simples, *Ann. Ec. Norm. Sup.*, *26* (1909), pp. 93–161.

[2] Cartan, H., and Eilenberg, S., Foundations of fibre bundles, *Symposium Internacional de Topología Algebraica, Mexico, D.F.*, 1958, pp. 16–23.

[3] Chern, S.-S., Pseudo-groupes continus infinis, *Colloque International de Géométrie Différentielle*, Strasbourg, (1953), pp. 119–136.

[4] Frölicher, A., and Nijenhuis, A., A theorem on stability of complex structures, *Proc. Nat. Acad. Sci., U.S.A. 43* (1957), pp. 239–241.

[5] Gray, J. W., Some global properties of contact structures, *Annals of Math.*, *69* (1959), pp. 421–450.

[6] Kodaira, K., and Spencer, D. C., On deformations of complex analytic structures, I-II, *Annals of Math.*, *67* (1958), pp. 328–466.

[7] Reeb, G., Sur certaines propiétés topologiques des variétés feuilletées, *Publications de l'Institut de Mathématiques de l'Université de Strasbourg*, *XI*, Hermann, Paris, 1952.

[8] Reinhart, B. L., Harmonic integrals on almost product manifolds, *Trans. Amer. Math. Soc.*, *88* (1958), pp. 243–276.

Quasiconformal Mappings and Teichmüller's Theorem†

Lipman Bers
NEW YORK UNIVERSITY

1. Introduction

LET S be a Riemann surface and f a homeomorphism of S onto another Riemann surface S'. If f is, in terms of local parameters, of class C_1 and has a positive Jacobian, then the deviation of this mapping from conformality can be measured, at each point p of S, by the ratio $K_p > 1$ of the axes of the infinitesimal ellipse into which f takes an infinitesimal circle located at p. We set $K[f] = \sup K_p$. The "overall dilatation" $K[f]$ can be defined also for a somewhat wider class of "quasiconformal" homeomorphisms.

Teichmüller's theorem asserts that given any orientation-preserving homeomorphism $f: S \to S'$ between two closed Riemann surfaces of genus $g > 1$, there exists, among all mappings homotopic to f, a unique "extremal" f_0 which minimizes $K[f]$. Furthermore, this extremal homeomorphism can be described analytically by two holomorphic quadratic differentials, $\phi(z)dz^2$ defined on S and $\psi(z')dz'^2$ defined on S'. In terms of local parameters $\zeta = \xi + i\eta$ and $\zeta' = \xi' + i\eta'$ such that $d\zeta^2 = \phi dz^2$, $d\zeta'^2 = \psi dz'^2$ the mapping f_0 can be written in the form: $\xi' = K\xi$, $\eta' = \eta$. Thus the extremal mapping is real analytic, except at the isolated zeros of the quadratic differentials, and the "local dilatation" K_p is a constant. Finally Teichmüller calls $\log K[f_0]$ the distance between the Riemann surfaces S and S', relative to the homotopy class of f, and asserts that the space of "topologically determined" or, as we shall say, "marked" Riemann surfaces, with this distance function is homeomorphic to the Euclidean space of $6g-6$ dimensions.

In his first memoir [22] of 1940, Teichmüller succeeded only in proving the uniqueness part of this theorem. Later [23] he gave an existence proof based on the continuity method. The idea of this proof is quite simple,

† Research leading to this paper was supported by the Office of Ordnance Research, United States Army, Contract No. DA-30-069-ORD-2153.

but the details are rather cumbersome. Another existence proof was found several years ago by Ahlfors [1].

In view of the importance and the beauty of Teichmüller's result† it seems worthwhile to give here a new version of the existence proof. Our arrangement of the argument preserves the logical structure of Teichmüller's proof; the details are carried out differently. More precisely, we work with the most general definition of quasiconformality, we rely on the theory of partial differential equations in some crucial parts of the argument, and we make use of a simple set of moduli for marked Riemann surfaces. In order to make this presentation self-contained, we shall also sketch Teichmüller's uniqueness proof.

Actually the Teichmüller theory refers to a more general situation. One considers not only closed Riemann surfaces, but also surfaces bounded by a finite number of analytic curves, and the homeomorphisms considered may be required to take on prescribed values at a finite number of prescribed points. But it is known (see Ahlfors' paper for details) that this more general situation can be reduced to the case of closed surfaces by means of doubling and by introducing appropriate two- or four-sheeted covering surfaces. Thus, for the sake of brevity, we shall be concerned only with closed surfaces.

2. Generalized derivatives

It is convenient to base the definition of quasiconformality on the concept of generalized derivatives due to Sobolev [21] and Friedrichs [7]. We recall the definition of these derivatives.

A. Let $f(x, y)$, $g(x, y)$, $h(x, y)$ be measurable functions§ defined in a domain D of the z-plane ($z = x + iy$) which belong to L_2 over every compact subset of D. One says that g and h are the generalized partial derivatives of f with respect to x and y respectively, and one writes $g = f_x$, $h = f_y$ if the following three conditions are satisfied:

(i) For every function ω of class C_1 which vanishes outside a compact set in D

$$\iint_D (f\omega_x + g\omega)dxdy = \iint_D (f\omega_y + h\omega)dxdy = 0.$$

(ii) For every compact subset $\Sigma \subset D$ there exists a sequence of functions $f^{(j)}$ of class C_1 such that

$$\iint_\Sigma (|f - f^{(j)}|^2 + |g - f_x^{(j)}|^2 + |h - f_y^{(j)}|^2)dxdy \to 0.$$

† Cf. Plutarch [19], "... it does not of necessity follow that, if the work delights you with its grace, the one who wrought it is worthy of your esteem".

§ Functions differing on a set of (two-dimensional) measure zero will be considered identical.

(iii) The function $f(x, y)$ is absolutely continuous in one variable for almost all (relevant) values of the other, and the ordinary partial derivatives of f are almost everywhere equal to g and h, respectively.

The basic theorem on generalized derivatives asserts that each of the conditions (i)–(iii) implies the other two. The implications (ii) $\rightarrow$ (iii) and (iii) $\rightarrow$ (i) are almost obvious. To prove the implication (i) $\rightarrow$ (ii) one uses the so-called mollifiers or some other smoothing device. Details will be found, for instance, in Smirnov's course [20].

B. The rules for operating with generalized derivatives follow easily from the definition and need not be spelled out. We mention explicitly only a lemma on convergence.

Assume that $\{f^{(j)}\}$ is a sequence of functions defined in a domain D and having generalized derivatives there. Assume also that the Dirichlet integrals of the functions $f^{(j)}$ are uniformly bounded and that the sequence $\{f^{(j)}\}$ converges in the L_2 sense. Then the limit function $f = \lim f^{(j)}$ has generalized derivatives which are weak limits, in the L_2 sense, of the corresponding generalized derivatives of a properly chosen subsequence $\{f^{(j_n)}\}$.

Indeed, local weak compactness of Hilbert space implies the existence of a subsequence $\{f^{(j_n)}\}$ such that the sequences $\{f_x^{(j_n)}\}$ and $\{f_y^{(j_n)}\}$ have weak limits g and h, and definition (i) of generalized derivatives shows that $g = f_x, h = f_y$.

C. The formal derivatives of a function $w(z)$ with respect to the complex variables $z = x + iy$ and $\bar{z} = x - iy$ are defined by the formulas

$$2w_z = w_x - iw_y, \quad 2w_{\bar{z}} = w_x + iw_y$$

These derivatives obey the usual rules of calculus and they may be understood either in the classical or in the generalized sense.

Let $w = u + iv$ be a complex-valued function defined in a domain D. The equation $w_{\bar{z}} = 0$ is equivalent to the Cauchy-Riemann system for the functions u, v. It is known and easy to verify that if the function w has generalized derivatives and $w_{\bar{z}} = 0$ almost everywhere, then w is an analytic function of z.

3. Quasiconformal mappings of plane domains

In this section we define the concept of quasiconformality for univalent functions in the plane. This definition will be extended to mappings of Riemann surfaces in §7.

A. Let $w = w(z)$ be a homeomorphism of a plane domain D, of class C_1 and with a positive Jacobian. Consider an infinitesimal circle located at some point z of D; the mapping w takes it into an infinitesimal ellipse.

The ratio $K(z)$ of the major axis of this ellipse to the minor axis measures the deviation of the mapping from conformality; we have

$$K = \frac{\max\limits_{0 \le \theta \le 2\pi} |\cos\theta \cdot w_x + \sin\theta \cdot w_y|}{\min\limits_{0 \le \theta \le 2\pi} |\cos\theta \cdot w_x + \sin\theta \cdot w_y|}$$

and, setting

$$K = \frac{1+k}{1-k},$$

also

$$k = \frac{|w_{\bar{z}}|}{|w_z|}.$$

The mapping is called quasiconformal if $K(z)$ is uniformly bounded, that is if there is a number $k < 1$ such that

$$|w_{\bar{z}}| \le k|w_z| \qquad (k < 1) \tag{3.1}$$

in D. Setting $w = u + iv$, this inequality may also be written as

$$u_x^2 + u_y^2 + v_x^2 + v_y^2 \le \left(K + \frac{1}{K}\right)(u_x v_y - u_y v_x). \tag{3.1'}$$

This definition of quasiconformality is, however, too restrictive. Thus Teichmüller permits the mapping to fail to be continuously differentiable at isolated points, and the most interesting quasiconformal mappings actually possess such singular points. A natural definition of quasiconformality, occurring for the first time in Morrey [16], is obtained by interpreting the derivatives in (3.1) as generalized derivatives. Thus we shall call a homeomorphism w quasiconformal (more precisely, k-quasiconformal) if w has generalized derivatives satisfying the inequality (3.1) almost everywhere.

B. Another definition of quasiconformality, free of all explicit differentiability requirements, is due to Ahlfors [1] and Pfluger [18]. A topological rectangle R with modulus m is, by definition, a conformal image of the Euclidean rectangle $0 \le \xi \le m, 0 \le \eta \le 1$. We write $m = \text{mod } R$. By virtue of Riemann's mapping theorem a topological image of a topological rectangle is again a topological rectangle. According to the "geometric" definition, a homeomorphism w of a plane domain D is k-quasiconformal if

$$\frac{\text{mod } w(R)}{\text{mod } R} \le \frac{1+k}{1-k} \tag{3.2}$$

for every topological rectangle $R \subset D$.

A classical result of Grötzsch [8, 9] shows that this inequality holds for a

C_1 mapping which is quasiconformal in the sense of A. It is also known that (3.2) holds for quasiconformal mappings with generalized derivatives. This can be seen for instance, by repeating Grötzsch's argument and using the result of §4, C. On the other hand, in view of a theorem by Mori [14], inequality (3.2) has as its consequences the existence of generalized derivatives and the differential inequality (3.1), as I have shown elsewhere [3]. The two definitions are thus equivalent.

C. A direct consequence of the geometric definition is the fact that the composite of a k_1-quasiconformal mapping with a k_2-quasiconformal one is k-quasiconformal, where

$$\frac{1+k}{1-k}=\frac{1+k_1}{1-k_1}\cdot\frac{1+k_2}{1-k_2}.$$

In particular, a k-quasiconformal mapping remains so if followed or preceded by a conformal transformation.

Another consequence of the geometric definition is the fact that the inverse mapping of a k-quasiconformal mapping is also k-quasiconformal.

D. One should mention another definition of quasiconformality which goes back to Lavrent'ev [11] (cf. Volkoviskii [24] and the references given there). Let w be a homeomorphism of D and set, for $z \in D$,

$$\hat{K}(z)=\limsup_{r\to 0}\frac{\max\limits_{0\le\theta\le 2\pi}|w(z+re^{i\theta})-w(z)|}{\min\limits_{0\le\theta\le 2\pi}|w(z+re^{i\theta})-w(z)|}.$$

According to the "differential-geometrical" definition w is called quasiconformal if $\hat{K}$ is bounded; it is called k-quasiconformal if

$$\hat{K}(z)\le\frac{1+k}{1-k}$$

almost everywhere. The equivalence of this definition with the previous two follows from the work of Pesin [17] (who considers a more general case) and Jenkins [10].

E. We state now the basic inequality for quasiconformal mappings. *Let $w(z)$ be a k-quasiconformal homeomorphism of the unit disc onto itself which leaves the origin fixed. There exist positive constants A, α depending only on k such that*

$$(3.3)\qquad |w(z_1)-w(z_2)|\le A|z_1-z_2|^{\alpha},\quad |z_1-z_2|\le A|w(z_1)-w(z_2)|^{\alpha}.$$

Thus every quasiconformal homeomorphism of the open unit disc onto itself is automatically a homeomorphism of the closed unit disc.

This result occurs first in Morrey [16]. The sharp exponent is $\alpha = (1-k)/(1+k)$ (Lavrent'ev [12], Ahlfors [1]). The best value of the constant A, independent of k, is 16 (Mori [15]).

Inequality (3.3) implies, of course, similar inequalities for k-quasiconformal mappings normalized in a different way, for instance, by prescribing the images of three points on the unit circle.

4. Beltrami equations

Our proof of Teichmüller's existence theorem will be based on the theory of Beltrami equations with bounded measurable coefficients. We collect here the necessary results, which are due to Morrey (cf. also Bers and Nirenberg [4], Boyarskiĭ [5]). Self-contained proofs will be found in a forthcoming paper by Ahlfors and Bers [2]. We remark that it would be easy, though somewhat awkward, to re-word our proof in such a way that it would involve only Beltrami equations with coefficients which are real analytic except at isolated points. For such equations the results stated below are classical.

A. By a Beltrami equation for a complex-valued function $w(z) = u + iv$ we mean an equation of the form

$$w_{\bar{z}} = \mu(z) w_z \tag{4.1}$$

where $\mu(z)$ is a measurable function satisfying the inequality

$$|\mu(z)| \leq k < 1. \tag{4.2}$$

By a solution we mean a continuous function having generalized derivatives which satisfy (4.1) almost everywhere.

We associate with μ the Riemannian metric

$$ds^2 = Edx^2 + 2Fdxdy + Gdy^2 = \lambda(z)|dz + \mu d\bar{z}|^2, \quad \lambda > 0$$

(and note that every Riemannian metric can be written in this form). Equation (4.1) expresses the fact that w is a mapping conformal with respect to this metric. One sees this by writing (4.1) in real form; the equations read

$$Wu_x = -Fv_x + Ev_y, \quad Wu_y = -Gv_x + Fv_y$$

where $W^2 = EG - F^2$.

We state now the main theorems on Beltrami equations.

B. *If $w_1(z)$ is a solution of the Beltrami equation* (4.1) *and $f(w)$ an analytic function, then $w_2(z) = f(w_1(z))$ is also a solution of* (4.1). *Conversely, if $w_1(z)$ and $w_2(z)$ are two solutions of the same Beltrami equation, defined in the same domain, and $w_1(z)$ is a homeomorphism, then $w_2(z) = f(w_1(z))$ where $f(w)$ is an analytic function.*

We remark that if $\mu(z)$ satisfies a Hölder condition, then all solutions of (4.1) are of class C_1, and for such classical solutions the assertion is, of course, trivial.

C. *A homeomorphic solution* $w(z) = u + iv$ *of a Beltrami equation takes (two-dimensionally) measurable sets into measurable sets, and for every such set* Δ

$$\iint_{\Delta} (u_x v_y - u_y v_x) dx dy = \iint_{w(\Delta)} du dv.$$

D. *Let* $\mu(z)$ *be a measurable function defined in the unit disc and satisfying* (4.2). *There exist solutions* $w = u + iv$ *of the Beltrami equation* (4.1) *which map the unit disc topologically onto itself.*

For every such solution we have, by (3.1′) and C.,

$$\text{(4.3)} \quad \iint_{|z|<1} (u_x^2 + u_y^2 + v_x^2 + v_y^2)^2 dx dy \leq 2\frac{1+k^2}{1-k^2} \iint_{|z|<1} (u_x v_y - u_y v_x) dx dy = 2\pi \frac{1+k^2}{1-k^2}.$$

It follows from B. that all solutions considered may be expressed in terms of a single solution w_0 by the formula

$$w(z) = e^{i\theta} \frac{w_0(z) - \zeta}{1 - \bar{\zeta} w_0(z)}$$

where θ is real and $|\zeta| < 1$.

E. The connection between Beltrami equations and quasiconformal mappings is contained in the following self-evident but important statement: Every homeomorphic solution of a Beltrami equation satisfying (4.2) is a k-quasiconformal mapping; conversely, every k-quasiconformal mapping is a solution of some Beltrami equation satisfying (4.2).

It follows from §3, E. that every solution considered in D. realizes a homeomorphism of the closed unit disc onto itself. Thus a solution may be determined uniquely by prescribing the images of three boundary points, or the images of one interior point and one boundary point.

F. We conclude by establishing the continuous dependence of the solutions discussed in D. on the coefficient μ.

Let there be given a sequence of measurable functions $\mu_j(z)$ *defined in the unit disc, and assume that*

$$\text{(4.4)} \qquad |\mu_j| \leq k < 1, \quad j = 1, 2, \ldots; \qquad \mu_j(z) \to \mu(z) \quad \text{a.e.}$$

Let $w^{(j)}(z)$ *denote the uniquely determined solution of the Beltrami equation* $w_{\bar{z}}^{(j)} = \mu_j w_z^{(j)}$ *which maps the unit disc topologically onto itself and satisfies the conditions*

$$\text{(4.5)} \qquad w^{(j)}(0) = 0, \; w^{(j)}(1) = 1.$$

Then the sequence $\{w^j\}$ *converges uniformly to the uniquely determined* *solution w of the Beltrami equation* (4.1) *which maps the unit disc onto itself leaving the points* 0, 1 *fixed.*

PROOF. The functions $w^{(j)}$ are all k-quasiconformal and it follows from §3, E. and Arzéla's theorem that a properly chosen subsequence $\{w^{(j_n)}\}$ converges uniformly to a function w which is also a homeomorphism of the closed unit disc onto itself leaving the points 0, 1 fixed. By D. the Dirichlet integrals of the functions $w^{(j)}$ are uniformly bounded. Hence, by §2, B., the function w has generalized derivatives and we may assume, selecting a subsequence if need be, that $w_x^{(j_n)} \to w_x$, $w_y^{(j_n)} \to w_y$ weakly, in the L_2 sense. Then also $w_z^{(j_n)} \to w_z$, $w_{\bar{z}}^{(j_n)} \to w_{\bar{z}}$ weakly, and by (4.4), $\mu_j w_z^{(j_n)} \to \mu w_z$ weakly. Hence w is a solution of (4.1). Thus w is uniquely determined (by E.), so that the selection of a subsequence $\{w^{(j_n)}\}$ was unnecessary.

REMARK. It is clear from the proof that the normalization condition (4.5) could be replaced by any other condition which normalizes a conformal mapping of the unit disc onto itself.

5. Representation of Riemann surfaces by Fuchsian groups

Before extending the definition of quasiconformality to mappings of Riemann surfaces we recall the representation of Riemann surfaces by fixed-point-free discontinuous groups of non-Euclidean motions. For the sake of brevity, such groups will be called simply Fuchsian groups.

A. We denote by U the upper half-plane of the complex z-plane. In U we introduce the Poincaré metric based on the line element $ds = |dz|y^{-1}$; this makes U into a model of the non-Euclidean plane. The geodesics of this metric (non-Euclidean straight lines) are arcs of Euclidean circles or straight lines orthogonal to the real axis. The non-Euclidean distance between two points will be denoted by $\langle z_1, z_2 \rangle$.

We denote by W the group of all homeomorphisms of U onto itself, by W_0 the subgroup of orientation-preserving homeomorphisms, and by W^* the subgroup of W consisting of mappings which are continuous and one-to-one on the extended real axis. The mapping

$$Q(z) = \frac{z - i}{z + i} \tag{5.1}$$

takes U into the unit disc. An element $T \in W$ belongs to W^* if and only if QTQ^{-1} is a homeomorphism of the closed unit disc.

The subgroup of W_0 which leaves the non-Euclidean distance invarians is the group $H \subset W^*$ of non-Euclidean motions. The elements of H are the linear transformations

$$T(z) = \frac{\alpha z + \beta}{\gamma z + \delta} \quad (\alpha, \beta, \gamma, \delta \text{ real}, \alpha\delta - \beta\gamma > 0).$$

By 1 we denote the identity transformation.

We topologize W by requiring $T_n \to T$ to mean that $T_n(z) \to T(z)$ uniformly on compact subsets of U. We topologize W^* by requiring $T_n \to T$ to mean that $QT_nQ^{-1} \to QTQ^{-1}$ uniformly on the closed unit disc. For elements of H both convergence concepts coincide.

B. An element $T \in H$ distinct from 1 is called a non-Euclidean translation if it leaves a non-Euclidean line (axis of the translation) invariant. An equivalent condition is that T leave exactly two points on the real axis fixed; these points are the end-points of the axis of T. Of the two fixed points one, which we denote by $\sigma = \sigma(T)$, is repelling, the other, denoted by $\tau = \tau(T)$, is attracting: $T^{-n}(z) \to \sigma$ and $T^n(z) \to \tau$ for $n = 1, 2, \ldots$. A translation may be written, uniquely, in the form

$$\frac{T(z) - \sigma}{T(z) - \tau} = \lambda \frac{z - \sigma}{z - \tau}, \quad \lambda > 1,$$

with an obvious interpretation of this formula in case either σ or τ is the point ∞. The number $\lambda = \lambda(T)$ is called the invariant of T.

C. By a Fuchsian group we shall mean here a subgroup $G \subset H$ such that 1 is an isolated point of G and $1 \neq T \in H$ implies that $T(z) \neq z$ for all $z \in U$. Let G be such a group. For $z \in U$ the equivalence class $[z]_G$ of z under G consists of all points $T(z)$, $T \in G$; it is known to be a discrete point set. Let U/G denote the set of equivalence classes; the natural mapping $\Pi_G : U \to U/G$ sends z into $[z]_G$. We make U/G into a Riemann surface by requiring Π_G to be, locally, a conformal homeomorphism.

Uniformization theory implies that every Riemann surface which is not the sphere, the sphere punctured at one or two points, or a closed surface of genus 1, can be represented as U/G. The representation is not unique, but U/G_1 and U/G_2 are conformally equivalent Riemann surfaces if and only if there is an $A \in H$ such that $G_1 = AG_2A^{-1}$.

D. A fundamental region of a Fuchsian group G is a set $\Sigma \subset U$ such that $\Pi_G(\Sigma) = U/G$ and the restriction of Π_G to the interior of Σ is one-to-one. The Riemann surface U/G is closed (compact) if and only if G has a compact fundamental region.

A Fuchsian group G with a compact fundamental region Σ consists of 1 *and of non-Euclidean translations.*

For the sake of completeness, we prove this well known result. Let $1 \neq T \in G$ and set $d = \inf \langle z, T(z) \rangle$, $z \in U$. For every $z \in U$ there is a $\zeta \in \Sigma$ and an $A \in G$ such that $z = A(\zeta)$, and $\langle A(\zeta), TA(\zeta) \rangle = \langle \zeta, A^{-1}TA(\zeta) \rangle$. Hence $d = \inf \langle \zeta, A^{-1}TA(\zeta) \rangle$, $\zeta \in \Sigma, A \in G$. Since Σ is compact and G discrete, there are elements A_0 and $T_0 = A_0^{-1}TA_0$ in G and a point ζ_0 such that $d = \langle \zeta_0, T_0(\zeta_0) \rangle$. Set $\zeta_1 = T_0(\zeta_0)$, $\zeta_2 = T_0^2(\zeta_0)$ and let z_0 be the midpoint of the non-Euclidean segment (ζ_0, ζ_1). Then $z_1 = T_0(z_0)$ is the midpoint of the non-Euclidean segment (ζ_1, ζ_2) and $\langle z_0, z_1 \rangle \leq \langle z_0, \zeta_1 \rangle +$

$\langle \zeta_1, z_1 \rangle = d$. On the other hand $\langle z_0, z_1 \rangle \geq d$. Thus $\langle z_0, z_1 \rangle = d$ and the three distinct points $\zeta_0, \zeta_1, \zeta_2$ lie on a non-Euclidean line l which is invariant under T_0. Hence T_0 is a translation and so is $T = A_0 T_0 A_0^{-1}$.

E. Let G be a Fuchsian group and $S = U/G$. By $\pi_1(S, p)$ we denote the fundamental group of S at a point $p \in S$. The elements of $\pi_1(S, p)$ are the homotopy classes $[\alpha]_p$ of closed curves α beginning and ending at p, and $[\alpha]_p[\beta]_p = [\alpha\beta]_p$. An isomorphism $\omega : \pi_1(S, p) \to \pi_1(S, q)$ will be called allowable if there is a curve γ on S leading from q to p such that $\omega([\alpha]_p) = [\gamma\alpha\gamma^{-1}]_q$. In particular, every inner automorphism of $\pi_1(S, p)$ is allowable.

For every $z \in U$ there is a natural isomorphism $\Lambda_{G,z}$ of G onto $\pi_1(U/G, \Pi_G(z))$ defined as follows.

Let $z_0 \in U$, $p_0 = \Pi_G(z_0)$. For $T \in G$ let α_T denote a curve in U leading from z_0 to $T(z_0)$ and set $\alpha'_T = \Pi_G(\alpha_T)$, $\Lambda_{G,z_0}(T) = [\alpha'_T]_{p_0}$. Then Λ_{G,z_0} is a canonical isomorphism of G onto $\pi_1(S, p_0)$.

LEMMA. *Let z_1 and z_2 be two points of U, $p_1 = \Pi_G(z_1)$, $p_2 = \Pi_G(z_2)$, and let $\lambda : G \to G$ and $\omega : \pi_1(S, p_1) \to \pi_1(S, p_2)$ be isomorphisms onto connected by the relation*

$$\lambda = \Lambda_{G,z_2}^{-1} \omega \Lambda_{G,z_1}.$$

Then λ is an inner automorphism if and only if ω is allowable.

The proof may be omitted.

6. Continuous mappings of Riemann surfaces

In this section we recall the representation of continuous mappings of Riemann surfaces by mappings of the upper half-plane. This is needed in order to apply the results on Beltrami equations.

A. Let S and S' be two Riemann surfaces. Two homomorphisms $\lambda : \pi_1(S, p) \to \pi_1(S', p')$ and $\nu : \pi_1(S, q) \to \pi_1(S', q')$ will be called equivalent if there exist allowable isomorphisms (cf. §5, E.) ω and ω' such that $\nu = \omega' \lambda \omega$.

Let $f : S \to S'$ be a continuous mapping. For every $p \in S$ it induces a homomorphism $f_p : \pi_1(S, p) \to \pi_1(S', f(p))$ defined by $f_p([\alpha]_p) = [f(\alpha)]_{f(p)}$. For any two points p and q of S the homomorphisms f_p and f_q are equivalent.

Two continuous mappings of S into S′ are homotopic if and only if they induce equivalent homomorphisms of the fundamental groups. We need not assume this known theorem since a proof, for the case of surfaces with a hyperbolic universal covering surface, will result from the following considerations.

B. Let G and G' be Fuchsian groups, $w : U \to U$ and $f : U/G \to U/G'$ continuous mappings and $\chi : G \to G'$ a homomorphism. We say that w

induces f if

$$f\Pi_G = \Pi_{G'} w. \tag{6.1}$$

We say that w induces χ if

$$wT = \chi(T)w \quad \text{for} \quad T \in G. \tag{6.2}$$

Note that, by (6.2), the element $\chi(T) \in G'$ is uniquely determined by T and w, since an element of G' is determined by what it does to a single point of U. Also, if (6.2) holds for some mapping $\chi : G \to G'$, this mapping must be a homomorphism, since we must have $\chi(T_1T_2)w = wT_1T_2 = \chi(T_1)wT_2 = \chi(T_1)\chi(T_2)w$. If w induces χ and $A \in G'$, then the mapping Aw induces the homomorphism $\chi' = A\chi A^{-1}$. If w induces f, then f is uniquely determined and is also induced by every mapping Aw, $A \in G'$.

C. *A continuous mapping $w : U \to U$ induces a continuous mapping $f : U/G \to U/G'$ if and only if it induces a homomorphism $\chi : G \to G'$. Every continuous mapping $f : U/G \to U/G'$ is induced by a mapping $w : U \to U$, and w is uniquely determined by f, except that it may be replaced by Aw, $A \in G'$.*

PROOF. Assume that w is given and (6.2) holds. Then the equivalence class $[w(z)]_{G'}$ depends only on the equivalence class $[z]_G$ so that (6.1) defines a mapping $f : U/G \to U/G'$. One verifies at once that the latter mapping is continuous. Assume next that (6.1) holds for some continuous mapping $f : U/G \to U/G'$. For every $z \in U$ and $T \in G$ there is a $\chi(z, T) \in G'$ such that $w(T(z)) = \chi(z, T)(w(z))$. But $\chi(z, T)$ must depend continuously on z: since G' is discrete, $\chi(z, T) = \chi(T)$. That χ is a homomorphism follows from B.

Now let $f : U/G \to U/G'$ be a given continuous mapping. For every z_0 in U there is an equivalence class $[\zeta_0]_{G'}$ such that $f(\Pi_G(z_0)) = [\zeta_0]_{G'}$. If we choose some element ζ_1 in $[\zeta_0]_{G'}$, then there exists a uniquely determined continuous function $w(z)$ defined in a neighborhood of z_0 such that $w(z_0) = \zeta_1$ and (6.1) holds. Since U is simply connected, the definition of w can be uniquely continued over the whole of U (monodromy theorem). The only free choice we had was that of the point ζ_1. Replacing ζ_1 by $\zeta_2 = A(\zeta_1)$, $A \in G'$, amounts to replacing w by Aw.

D. *Let $w_0 : U \to U$ and $w_1 : U \to U$ be continuous mappings which induce the homomorphisms $\chi_0 : G \to G'$ and $\chi_1 : G \to G'$, and the mappings $f_0 : U/G \to U/G'$ and $f_1 : U/G \to U/G'$, respectively. The mappings f_0 and f_1 are homotopic if and only if there is an $A \in G'$ such that $\chi_1 = A\chi_0A^{-1}$.*

PROOF. (following Ahlfors). Assume first that $\chi_1 = A\chi_0A^{-1}$, $A \in G'$. Replacing w_1 by $A^{-1}w_1$ we may assume that $\chi_1 = \chi_0$. For every t, $0 \leq t \leq 1$, let the point $w_t(z)$ divide the non-Euclidean segment $(w_0(z), w_1(z))$ in the ratio $t : 1 - t$. One verifies that $w_t(z)$ depends continuously on t and z, and that $w_tT = \chi_0(T)w_t$ for all $T \in G$. Hence (cf. C.) w_t induces a continuous mapping $g_t : U/G \to U/G'$. This mapping depends continuously on t and $g_0 = f_0$, $g_1 = f_1$.

Assume now that the mappings f_0 and f_1 are homotopic so that there exists a continuous mapping $g_t : U/G \to U/G'$ depending continuously on t, $0 \leq t \leq 1$ with $g_0 = f_0$, $g_1 = f_1$. Choose points z_0, ζ_0 so that $f(\Pi_G(z_0)) = \Pi_{G'}(\zeta_0)$. There is a continuous complex-valued function $\delta(t)$, $0 \leq t \leq 1$, with $\delta(0) = \zeta_0$ and $g_t(\Pi_G(z_0)) = \Pi_{G'}(\delta(t))$. By the proof of C. there exists a continuous mapping $\hat{w}_t : U \to U$ satisfying the condition $w_t(z_0) = \delta(t)$ and inducing g_t. This mapping is easily seen to depend continuously on t. Let $\hat{\chi}_t : G \to G'$ be the homomorphism induced by $\hat{w}_t$. For every $T \in G$ the element $\hat{\chi}_t(T) \in G'$ depends continuously on t; since G' is discrete, $\hat{\chi}_t = \hat{\chi}_0$. By C. we have that $w_0 = A_0\hat{w}_0$, $w_1 = A_1\hat{w}_1$, where $A_0, A_1 \in G'$. Hence $\chi_0 = A_0\hat{\chi}_0A_0^{-1}$, $\chi_1 = A_1\hat{\chi}_0A_1^{-1}$, and setting $A = A_1A_0^{-1}$ we have $\chi_1 = A\chi_0A^{-1}$.

E. *Let the continuous mapping $w : U \to U$ induce the continuous mapping $f : U/G \to U/G'$ and the homomorphism $\chi : G \to G'$. The homomorphisms of fundamental groups $f_{\Pi_G(z_0)}$* and *$\Lambda_{G',w(z_0)}\chi\Lambda_{G,z_0}^{-1}$ are equivalent.*

The proof follows simply by recalling the definitions of f_p and $\Lambda_{G,z}$ given in the previous section. In conjunction with C. and D. we obtain a proof of the topological theorem stated in A.

F. *Under the hypothesis of* E. *the mapping f is a homeomorphism onto if and only if w is (that is, $w \in W$) and in this case χ is an isomorphism onto.*

Assume that f is a homeomorphism and $f(U/G) = U/G'$. Then f_p is clearly an isomorphism onto and the assertion concerning χ follows from E. Suppose that $w(z_1) = w(z_2)$. Then $f(\Pi_G(z_1)) = f(\Pi_G(z_2))$ so that there is a $T \in G$ with $z_2 = T(z_1)$, and by (6.1) we have $w(z_2) = \chi(T)(w(z_1)) = w(z_1)$. Since the group G' is fixed-point-free, $\chi(T) = 1$ and also $T = 1$, $z_2 = z_1$. Thus w is one-to-one. Finally, $w(U) = U$; for, if w would omit a value $\zeta \in U$ it would also omit all values in $[\zeta]_{G'}$ and f would omit the point $\Pi_{G'}(\zeta)$. It is plain that f is onto if w is and that w must fail to be one-to-one if f does.

We note that an orientation-preserving homeomorphism f is induced by an element of W_0.

G. An element $w \in W$ will be said to be compatible with the Fuchsian group G if $wGw^{-1} \subset H$.

If $w \in W$ is compatible with the Fuchsian group G, then $G' = wGw^{-1}$ is a Fuchsian group and w induces a homeomorphism of U/G onto U/G'.

PROOF. Every inner automorphism of W is a homeomorphism; hence G' is a discrete group of non-Euclidean motions. Assume that $A \in G'$ and $A(\zeta) = \zeta$ for some $\zeta \in U$. Then $T = w^{-1}Aw \in G$ and for $z = w^{-1}(\zeta)$ we have that $T(z) = z$. Hence $T = 1$, $A = 1$. Thus G' is fixed-point-free. Clearly w induces the isomorphism $\chi(T) = wTw^{-1}$ of G onto G'. The remaining statement follows from F.

7. Quasiconformal mappings of Riemann surfaces

In this section we define quasiconformality for mappings of Riemann surfaces and establish a one-to-one correspondence between quasiconformal homeomorphisms of a Riemann surface and a class of differentials defined on the surface.

A. Let S and S' be Riemann surfaces and $f : S \to S'$ a homeomorphism onto. Let z and ζ be local parameters defined in neighborhoods of points $p_0 \in S$ and $q_0 = f(p_0)$, respectively. Thus $z = g(p)$ and $\zeta = h(q)$ are conformal homeomorphisms of neighborhoods of the points p_0 and q_0 into the complex plane. Near p_0 the mapping f can be written in the form $\zeta = w(z)$. More precisely: $w = hfg^{-1}$. If this mapping w is k-quasiconformal (cf. §3, A.) we say that f is k-quasiconformal in the neighborhood of p_0. The definition is obviously independent of the choice of local parameters (cf. §3, C.). If f is k-quasiconformal in the neighborhood of every point of S, f is called quasiconformal, more precisely k-quasiconformal. We denote by $k[f]$ the smallest number k' such that f is k'-quasiconformal, and set $K[f] = (1 + k[f])/(1 - k[f])$. If f is not quasiconformal, we set $k[f] = 1$, $K[f] = \infty$.

It follows from §3, C. that $k[f^{-1}] = k[f]$ and that if $f_1 : S' \to S''$ is another homeomorphism, then

$$K[f_1 f] \leq K[f_1]K[f]. \tag{7.1}$$

0-quasiconformal mappings are conformal (cf. §2, C.).

B. A differential of type (m, n) on a Riemann surface S is a rule associating with every local parameter z defined on a domain $D \subset S$ a measurable function $\psi(z)$, $z \in D$, in such a way that the expression $\psi(z)dz^m d\bar{z}^n$ is invariant under parameter changes. For a differential $\mu(z)d\bar{z}/dz$ of type $(-1, 1)$ the absolute value of the coefficient, $|\mu(z)|$, is a scalar on S. If $|\mu| \leq k < 1$ we call $\mu(z)$ a Beltrami coefficient and $\mu d\bar{z}/dz$ a Beltrami differential.

There exists a natural one-to-one correspondence between Beltrami differentials on S and quasiconformal homeomorphisms of S.

Let $f : S \to S'$ be a given quasiconformal homeomorphism. Let z and $\zeta = w(z)$ have the same meaning as in A. The function $w(z)$ satisfies, in a neighborhood of $z_0 = g(p_0)$, a definite Beltrami equation $w_{\bar{z}} = \mu(z)w_z$ (cf. 4, D.). A direct computation (which is legitimate though we work with generalized derivatives) shows that $\mu d\bar{z}/dz$ is a differential of type $(-1, 1)$. We call it the Beltrami differential of the mapping f. Now let $\mu d\bar{z}/dz$ be a given Beltrami differential on S. We define on S a new conformal structure based on the conformal metric $|dz + \mu d\bar{z}|$ and call the new Riemann surface thus obtained S^μ. More precisely, the points and the open sets on S^μ are those of S and a local parameter on S^μ is a univalent function $\zeta = w(z)$ of a local parameter z of S which satisfies the Beltrami equation $w_{\bar{z}} = \mu w_z$. The legitimacy of this definition follows from the results stated in §4, B. and C. The identity mapping $f^\mu : S \to S^\mu$ is quasiconformal and its Beltrami coefficient is precisely μ; we call f^μ the natural mapping induced by $\mu d\bar{z}/dz$. It is clear that every quasiconformal mapping may be considered as a natural mapping induced by its Beltrami differential.

We norm the space of Beltrami differentials on S by setting $\|\mu d\bar{z}/dz\| = \|\mu\| =$ true maximum of $|\mu|$. Then the natural mapping induced by $\mu d\bar{z}/dz$ satisfies the equation $k[f^\mu] = \|\mu\|$.

C. Let G be a Fuchsian group. Since $z \in U$ is a local parameter in the neighborhood of every point of U/G we conclude (cf. §6, F.) that a homeomorphism $f: U/G \to U/G'$ is k quasiconformal if and only if it is induced by a k quasiconformal homeomorphism $w : U \to U$.

Every Beltrami differential on U/G may be written in the form $\mu(z)d\bar{z}/dz$ where $\mu(z)$ is a bounded measurable function defined in U and such that $|\mu| \leq k < 1$ and

$$\mu(T(z)) = \mu(z)T'(z)/\overline{T'(z)} \text{ for } T \in G. \tag{7.2}$$

We call such functions Beltrami coefficients compatible with G. For every Beltrami coefficient μ compatible with G, $\mu d\bar{z}/dz$ is a Beltrami differential on U/G.

D. *Let G be a Fuchsian group and $\mu(z)$ a Beltrami coefficient compatible with G. There exists a solution w of the Beltrami equation*

$$w_{\bar{z}} = \mu(z)w_z$$

which maps U topologically onto itself, and $w \in W_0 \cap W^*$. Moreover, w is compatible with G (cf. §6, G.), *$U/(wGw^{-1}) = (U/G)^\mu$, and w induces the natural mapping f^μ of U/G onto $(U/G)^\mu$.*

PROOF. Since the upper half-plane is conformally equivalent to the unit disc, the equivalence being established, say by the mapping Q (§5, A.), the existence of w follows from §4, B. and C. All solutions of Beltrami equations being orientation preserving, $w \in W_0$. That $w \in W^*$ follows from §3, E. Now set, for some $T \in G$, $\hat{w}(z) = w(T(z))$. A direct computation based on (7.2) shows that $\hat{w}_{\bar{z}} = \mu(z)\hat{w}_z$. (This computation is legitimate, though we work with generalized derivatives.) By §4, B. we conclude that there exists an analytic function $A(w)$ such that $\hat{w}(z) = w(T(z)) = A(w(z))$. But $A = wTw^{-1} \in W_0$, so that A is a non-Euclidean motion. Hence $wGw^{-1} \subset H$ and w is compatible with G. It follows that w induces a homeomorphism of U/G onto $U/(wGw^{-1})$. The proof of the remaining statements is obvious.

We remark that w is not uniquely determined by μ but may be replaced by $w_1 = Aw$, $A \in H$ (cf. §6, C.). But $w_1 G w_1^{-1} = A(wGw^{-1})A^{-1}$ so that the Riemann surfaces $U/(wGw^{-1})$ and $U/(w_1 G w_1^{-1})$ are conformally equivalent.

8. Teichmüller mappings

We shall define Teichmüller mappings using everywhere regular quadratic differentials, since such mappings turn out to be extremal in the problems considered here. Mappings defined by meromorphic quadratic differentials with only simple poles are extremal in the class of mappings attaining preassigned values at given points. It is not known what extremal property, if any, is enjoyed by mappings defined by meromorphic quadratic differentials with poles of higher order.

A. A quadratic differential on a Riemann surface S is a differential $\phi(z)dz^2$ of type (2, 0); it is called regular if the coefficient $\phi(z)$ is an analytic function of the local parameter z. Let such a differential be given, and assume that it does not vanish identically. At a point p of S we have then $\phi dz^2 = z^m(a_0 + a_1 z + \dots)dz^2$ where z is a local parameter which vanishes at p, and $a_0 \neq 0$. The non-negative integer $m = m(p)$ is called the order of the differential at p; if $m > 0$, p is called a zero of multiplicity m. Set

$$\zeta = \zeta(z) = \left\{\int_0^z \phi(z)^{1/2} dz\right\}^{2/(m+2)}. \tag{8.1}$$

Then ζ is a local parameter defined near p; we call it the natural parameter belonging to ϕdz^2 at p. The natural parameter is characterized by the relations.

$$\zeta = 0 \text{ at } p, \quad \phi dz^2 = d\zeta^2 \quad \text{if} \quad m = 0, \tag{8.2}$$

$$\phi dz^2 = \left(\frac{m+2}{2}\right)^2 \zeta^m d\zeta^2 \quad \text{if} \quad m > 0.$$

It may be multiplied by an $(m + 2)$nd root of unity, but is otherwise uniquely determined.

B. Let G be a Fuchsian group and $S = U/G$. Since $z \in U$ is a local parameter in the neighborhood of every point on S, every regular quadratic differential on S may be written as $\phi(z)dz^2$ where $\phi(z)$ is holomorphic in U and satisfies the functional equation

$$\phi(T(z)) = \frac{\phi(z)}{T'(z)^2} \quad \text{for all} \quad T \in G. \tag{8.3}$$

Conversely, if $\phi(z)$ is a holomorphic solution of (8.3) defined in U, then ϕdz^2 is a regular quadratic differential on S.

C. Let f be a homeomorphism of a Riemann surface S onto another such surface S'. We say that f is a Teichmüller mapping if f is conformal or if f is quasiconformal with a Beltrami coefficient (cf. §7, B.) of the form

$$\mu(z) = k\frac{\overline{\phi(z)}}{|\phi(z)|} \tag{8.4}$$

where $\phi dz^2 \not\equiv 0$ is a regular quadratic differential on S and k a constant such that $0 < k < 1$. In the latter case we say that f is defined by ϕdz^2 and k. It is easy to see that in this case f determines uniquely the differential ϕdz^2, except for a positive constant factor, and also that $K[f] = (1 + k)/(1 - k)$.

We remark that whenever $\phi(z)dz^2$ is a quadratic differential (regular or not) on S and $0 < k < 1$, (8.4) is a Beltrami coefficient on S.

D. *Let $f : S \rightarrow S'$ be a Teichmüller mapping defined by the regular quadratic differential $\phi(z)dz^2 \not\equiv 0$ on S and the constant k. Then there exists a uniquely determined regular quadratic differential $\psi(z')dz'^2$ on S' having the following properties.* (i) *The order of $\psi dz'^2$ at $f(p)$ is equal to the order of ϕdz^2 at p.* (ii) *Let ζ be the natural parameter belonging to ϕdz^2 at a point $p \in S$ at which ϕdz^2 has order zero. In the neighborhood of p the mapping f may be described by the equation*

$$\zeta' = \frac{\zeta + k\bar{\zeta}}{1 - k} \tag{8.5}$$

where ζ' is the natural parameter belonging to $\psi dz'^2$ at $f(p)$. (iii) *The inverse mapping $f^{-1} : S' \rightarrow S$ is a Teichmüller mapping defined by the differential $-\psi dz'^2$ and the constant k.*

Before proving this statement we remark that, setting $\zeta = \xi + i\eta$, $\zeta' = \xi' + i\eta'$, relation (8.5) may be written in the form

$$\xi' = K\xi, \quad \eta' = \eta \qquad (K = K[f]). \tag{8.6}$$

Thus a Teichmüller mapping is, in the neighborhood of every point which is not a zero of the defining quadratic differential, a conformal transformation, followed by a fixed affine transformation, followed by another conformal transformation. This itself already suggests the extremal nature of Teichmüller mappings, if one thinks of the simple case of mapping one topological rectangle onto another (cf. §3, B.).

E. In order to prove D. we consider some point $p \in S$ at which ϕdz^2 has order m. Let ζ be a natural parameter belonging to ϕdz^2 at p and set

$$\begin{aligned}\zeta' &= \left(\frac{\zeta^{(m+2)/2} + k\bar{\zeta}^{(m+2)/2}}{1 - k}\right)^{2/(m+2)} \\ &= (1 - k)^{-2/(m+2)}\zeta\sum_{j=0}^{\infty}\binom{2/(m+2)}{j}k^j\left(\frac{\bar{\zeta}}{\zeta}\right)^{j(m+2)/2}.\end{aligned} \tag{8.7}$$

If $m = 0$ this is relation (8.5).

Now let z' be some local parameter on S' defined near the point $f(p)$. In the neighborhood of p the mapping f is described by a function $z' = w(\zeta)$ which is a homeomorphic solution of the Beltrami equation

$$w_{\bar{\zeta}} = k\left(\frac{\bar{\zeta}}{|\zeta|}\right)^m w_{\zeta}. \tag{8.8}$$

But the function ζ' defined above is, for small values of $|\zeta|$, also a homeomorphic solution of this Beltrami equation. Thus ζ' is, by §4, B., an analytic function of $z' = w(\zeta)$ and hence a local parameter on S near $f(p)$. We call it, for the time being, the distinguished parameter at $f(p)$.

Now let $p_1 \neq p$ be a point sufficiently close to p and t the value of ζ at p_1. Since the zeros of a regular quadratic differential are isolated, the order of ϕdz^2 at p_1 is 0 and, by A., a natural parameter at p_1 is

$$\zeta_1 = \int_t^{\zeta} \left(\frac{m+2}{2}\right) \zeta^{m/2} d\zeta = \zeta^{(m+2)/2}.$$

Hence

$$\zeta_1' = \frac{\zeta_1 - k\bar{\zeta}_1}{1-k}$$

is the distinguished parameter on S' at $f(p_1)$, and comparing this with (8.7) we see that $4d\zeta_1'^2 = (m+2)^2 \zeta'^m d\zeta'^2$. This means that we may define a regular quadratic differential on S' by setting

$$\psi(z')dz'^2 = \left(\frac{m+2}{2}\right)^2 \zeta'^m d\zeta'^2.$$

This differential has properties (i) and (ii) by construction. Solving (8.7) for ζ we see that ζ as a function of ζ' satisfies the Beltrami equation

$$\zeta_{\bar{\zeta}'} = -k\left(\frac{\bar{\zeta}'}{|\zeta'|}\right)^m \zeta_{\zeta'}.$$

Consequently, $\psi dz'^2$ also has property (iii).

9. Marked Riemann surfaces

In this section we define marked Riemann surfaces, a concept which is identical in content with Teichmüller's topologically determined Riemann surfaces.

A. Let S be a closed Riemann surface of genus $g > 1$. A standard set of generators at a point $p \in S$ is an ordered set $a = (a_1, a_2, \ldots, a_{2g})$ of elements of the fundamental group $\pi_1(S, p)$ which generate this group with the single defining relation

$$\prod_{j=1}^{g} (a_{2j} a_{2j-1}^{-1} a_{2j}^{-1} a_{2j-1}) = 1.$$

The existence of a standard set is classical; the actual construction of such a set involves the so called canonical dissection of the surface.

Two standard sets of generators, a at p and b at q, will be called equivalent if there exists an allowable isomorphism (cf. §5, E.) $\omega : \pi_1(S, p) \to \pi_1(S, q)$ such that $\omega(a) = b$, that is $\omega(a_j) = b_j$, $j = 1, 2, \ldots, 2g$. A Riemann surface S together with an equivalence class α of standard sets of generators will be called a marked Riemann surface and will be denoted by (S, α).

An equivalent definition reads as follows. A marked Riemann surface is a closed Riemann surface S of genus $g > 1$ together with a canonical dissection of S; two dissections define the same marked surface, if one is obtained from the other by a homeomorphism of S onto itself which is homotopic to the identity.

B. Let (S, α) and (S', α') be two marked Riemann surfaces of the same genus. A homeomorphism $f : S \to S'$ will be called a mapping of (S, α) onto (S', α') and will be denoted by $f : (S, \alpha) \to (S', \alpha')$ if for a point $p \in S$ the induced isomorphism $f_p : \pi_1(S, p) \to \pi_1(S', f(p))$ takes a set of generators belonging to α into a set of generators belonging to α'.

If $f : (S, \alpha) \to (S', \alpha')$ and $h : (S, \alpha) \to (S', \alpha')$, then the mappings f and h are homotopic. Conversely, if $h : S \to S'$ is a homeomorphism homotopic to f, then $h : (S, \alpha) \to (S', \alpha')$.

The proof follows at once from §6, A.

Two Riemann surfaces are considered identical if they are conformally equivalent. Therefore two marked Riemann surfaces, (S, α) and (S', α') are considered *identical* if there exists a conformal homeomorphism $f : (S, \alpha) \to (S', \alpha')$.

C. *If (S, α) and (S', α') are two marked Riemann surfaces of the same genus, then there exists a homeomorphism $f : S \to S'$ such that $f : (S, \alpha) \to (S', \alpha')$.*

This follows from the known topological theorem (Dehn) which asserts that every automorphism of the fundamental group of a closed surface can be realized by a homeomorphism of the surface onto itself. For the proof we refer to the paper of Mangler [13] and the references given there.

The homeomorphism f is, of course, determined only up to a homotopy. If f is orientation-preserving (orientation-reversing), we say that the marked Riemann surfaces (S, α) and (S', α') are similarly oriented (have opposite orientations).

D. *If (S, α) and (S', α') are two similarly oriented marked Riemann surfaces of the same genus, then there exists a quasiconformal mapping $f : (S, \alpha) \to (S', \alpha')$.*

Had we demanded that the homeomorphism f be continuously differentiable everywhere, the proof would be somewhat laborious. Since we use a very general definition of quasiconformality, the proof presents no difficulties and may be omitted.

The assertion means that every homeomorphism $f: S \rightarrow S'$ is homotopic to a quasiconformal mapping. The corresponding assertion for open surfaces is trivially false. For instance, there is no quasiconformal mapping of the finite plane onto the unit disc.

10. Statement of Teichmüller's theorem

For the sake of convenience, we separate Teichmüller's theorem into three assertions.

A. THEOREM I. *Let* $f_0 : (S, \alpha) \rightarrow (S', \alpha')$ *be a Teichmüller mapping of one marked Riemann surface of genus* $g > 1$ *onto another such surface. If* $f : (S, \alpha) \rightarrow (S', \alpha')$ *is a mapping distinct from* f_0 *then* $K[f_0] < K[f]$.

The theorem states that for a closed Riemann surface of genus $g > 1$ a Teichmüller mapping has a strictly smaller dilatation than any homeomorphism homotopic to it. As a corollary we find that no two distinct Teichmüller mappings are homotopic.

B. THEOREM II. *Let* (S, α) *and* (S', α') *be two similarly oriented marked Riemann surfaces of the same genus* $g > 1$. *Then there exists a Teichmüller mapping* $f_0 : (S, \alpha) \rightarrow (S', \alpha')$.

In other words, every homotopy class of an orientation preserving homeomorphism $f: S \rightarrow S'$ contains a Teichmüller mapping. By the uniqueness theorem (Theorem I) this mapping is extremal.

C. The Teichmüller distance between two marked similarly oriented Riemann surfaces of genus $g > 1$ is, by definition,

$$d((S, \alpha), (S', \alpha')) = \log K[f_0] \tag{10.1}$$

where $f_0 : (S, \alpha) \rightarrow (S', \alpha')$ is the Teichmüller mapping which exists and is unique by virtue of I and II. An equivalent definition is

$$d((S, \alpha), (S', \alpha')) = \inf \log K[f] \tag{10.2}$$

taken over all mappings $f : (S, \alpha) \rightarrow (S', \alpha')$. The fact that the distance between two distinct marked Riemann surfaces is positive and that the distance function is symmetric is an immediate consequence of the definition. To prove the triangle inequality consider a Teichmüller mapping $f_1 : (S', \alpha') \rightarrow (S'', \alpha'')$. Using definition (10.2) and inequality (7.1), we conclude that the distance between (S, α) and (S'', α'') does not exceed $\log K[f_0] + \log K[f_1]$. Thus similarly oriented marked Riemann surfaces of genus $g > 1$ form a metric space R_g.

THEOREM III. *The metric space* R_g *is homeomorphic to the Euclidean space of* $6g-6$ *dimensions.*

This is a partial solution of the so-called problem of moduli.

11. The metric induced by a regular quadratic differential

This and the following section contain an outline of the proof of the uniqueness theorem (Theorem I). We merely re-word Teichmüller's own argument, making sure that it applies also to the general definition of quasiconformality used here. (Another modification of Teichmüller's uniqueness proof was given by Ahlfors.)

A. Let $\phi(z)dz^2$ be a regular quadratic differential defined on a closed Riemann surface S. We define on S a metric based on the line element $ds_\phi = |\phi(z)|^{1/2}|dz|$; the corresponding area element is $dA_\phi = |\phi(z)|dxdy$. The length of a curve α in this metric will be denoted by $|\alpha|_\phi$.

At points distinct from the zeros of the differential the metric is locally Euclidean; the zeros are singular points. A geodesic is characterized by the property that along it arg $\phi(z)dz^2$ is constant. A geodesic which passes through a zero of order m may possess a corner there, but with an angle not less than $2\pi/(m+2)$. One verifies this by using natural parameters. Hence, every geodesic may be continued indefinitely and to an arbitrarily given length, though the continuation through a zero is not unique. One now concludes, in the usual way, that if α is a curve on S with endpoints p and q, then the homotopy class of α relative to the endpoints, $[\alpha]_{p,q}$, contains a geodesic (which minimizes the ϕ-length). Furthermore, the geodesic is unique. This can be proved by repeating the classical argument for a surface with a smooth Riemannian metric of non-positive curvature. In fact, the metric ds_ϕ has a non-positive curvature in the sense that the sum of angles in a geodesic triangle does not exceed π.

B. A geodesic arc along which $\phi dz^2 > 0$ (or $\phi dz^2 < 0$) is called horizontal (vertical). A geodesic arc which does not pass through a zero of ϕdz^2 is called free. A free horizontal arc is the unique shortest curve in its homotopy class.

Let p_0 be a point which is not a zero of ϕdz^2, and let p_0 be the midpoint of a free vertical arc β of length $2b$. Assume further that every point of β is a midpoint of a free horizontal arc of length $2a$. Then these horizontal arcs cover a simply connected domain $\Delta(p_0, a, b)$ on the universal covering surface of S and the natural parameter $\zeta(p) = \xi + i\eta$ belonging to ϕdz^2 at p_0 maps $\Delta(p_0, a, b)$ conformally onto the rectangle $-a < \xi < a$, $-b < \eta < b$. This construction makes it clear what we mean by saying that some property holds for almost all horizontal arcs.

For $c > 0$ let Ω_c denote the set of those points of S which can be joined to a zero of ϕdz^2 by a horizontal arc of length not exceeding c. Since only a finite number of horizontal arcs emerge from each of the isolated zeros of ϕdz^2, the set Ω_c is the union of at most countably many analytic arcs (the union of finitely many analytic arcs if S is closed). The union Ω_∞ of all Ω_c has two-dimensional measure zero.

If p_0 is a point in the complement of Ω_c and $0 < a < c$, then there is a $b > 0$ such that we may form the domain $\Delta(p_0, a, b)$.

C. LEMMA A. *Let $\phi(z)dz^2 \not\equiv 0$ be a regular quadratic differential on a closed Riemann surface and let $f: S \to S$ be a homeomorphism homotopic to the identity. There exists a constant $A > 0$ such that for every free horizontal arc α*

$$|f(\alpha)|_\phi \geq |\alpha|_\phi - 2A. \tag{11.1}$$

PROOF. By assumption there exists a continuous mapping $f_t : S \to S$ depending continuously on t, $0 \leq t \leq 1$, with $f_1 = f$ and $f_0(p) = p$. For every $p_0 \in S$ let $\gamma(p_0)$ denote the curve $p = f_t(p_0)$, $0 \leq t \leq 1$. Let $\delta(p_0)$ be the unique geodesic in the homotopy class of $\gamma(p_0)$ and set

$$A = \sup |\delta(p)|_\phi, \qquad p \in S. \tag{11.2}$$

Compactness of S implies easily that $A < \infty$. Now let α be a free horizontal arc with endpoints p and q. The curve $\gamma(p)f(\alpha)\gamma(q)^{-1}$ has the same endpoints as α and is homotopic to it. The same is true of the curve $\delta(p)f(\alpha)\delta(q)^{-1}$. Thus $|\alpha|_\phi \leq |\delta(p)|_\phi + |f(\alpha)|_\phi + |\delta(q)|_\phi$ and comparing this inequality with (11.2) we obtain (11.1).

D. Let $\phi(z)dz^2 \not\equiv 0$ be a regular quadratic differential on a Riemann surface S, $\psi(z')dz'^2$ another such differential on a Riemann surface S' and f a quasiconformal homeomorphism of S onto S'. On S we define the measurable non-negative point-functions

$$\lambda_{f;\,\phi,\psi}(p) = |\phi(z)^{-1/2}w_z + \overline{\phi(z)}^{-1/2}w_{\bar{z}}|\ |\psi(w(z))|^{1/2}, \tag{11.3}$$

$$j_{f;\,\phi,\psi}(p) = |\phi(z)|^{-1}(|w_z|^2 - |w_{\bar{z}}|^2)\ |\psi(w(z))|. \tag{11.4}$$

Here z is some local parameter defined near $p_0 \in S$, w some local parameter defined near $f(p_0) \in S'$, and $w = w(z)$ the function defined by the mapping f. If p_0 and $f(p_0)$ are not zeros of the corresponding quadratic differentials we have, in terms of natural parameters $\zeta = \xi + i\eta$, $\zeta' = \xi' + i\eta'$,

$$\lambda_{f;\,\phi,\psi} = [(\xi'_\xi)^2 + (\eta'_\xi)^2]^{1/2}, \tag{11.3'}$$

$$j_{f;\,\phi,\psi} = \xi'_\xi\eta'_\eta - \xi'_\eta\eta'_\xi. \tag{11.4'}$$

This shows the simple geometric meaning of the functions defined. If $G \subset S$ is an open set, then

$$\iint_G j_{f;\,\phi,\psi}\,dA_\phi = \iint_{f(G)} dA_\psi, \tag{11.5}$$

and for almost all horizontal arcs α on S we have

$$\int_\alpha \lambda_{f;\,\phi,\psi}\,ds_\phi = \int_{f(\alpha)} ds_\psi = |f(\alpha)|_\psi. \tag{11.6}$$

We also note the inequality

$$\lambda^2_{f;\,\phi,\psi} \leq K[f]j_{f;\,\phi,\psi} \tag{11.7}$$

which holds almost everywhere and is an immediate consequence of the definition of quasiconformality.

E. The uniqueness proof is based on the following:

LEMMA B. *Let $\phi(z)dz^2 \not\equiv 0$ be a regular quadratic differential on a closed Riemann surface S and $f : S \rightarrow S$ a homeomorphism homotopic to the identity. Then*

$$\iint_S \lambda_{f;\,\phi,\phi} dA_\phi \geq \iint_S dA_\phi. \tag{11.8}$$

PROOF. Let $\nu(p)$ be any non-negative measurable integrable function defined on S. For some $a > 0$ define a new function $\tilde{\nu}(p, a)$ by setting

$$\tilde{\nu}(p, a) = \int_\alpha \nu(p) ds_\phi$$

if p is the midpoint of a free horizontal arc α with $|\alpha|_\phi = 2a$ and $\nu(p)$ is measurable and integrable on α, and $\tilde{\nu}(p, a) = 0$ otherwise. A simple argument, given below, shows that

$$\iint_S \tilde{\nu}(p, a) dA_\phi = 2a \iint_S \nu(p) dA_\phi. \tag{11.9}$$

We apply this relation to the function $\lambda_{f;\,\phi,\phi}$. Noting (11.6) and Lemma A we have that $\tilde{\lambda}_{f;\,\phi,\phi}(p, a) \geq 2(a - A)$ almost everywhere. Thus

$$2a \iint_S \lambda_{f;\,\phi,\phi} dA_\phi \geq 2(a - A) \iint_S dA_\phi,$$

and since this must hold for all $a > 0$, and A is fixed, inequality (11.8) follows.

F. It remains to verify relation (11.9). It is clear that we may assume that the function ν is continuous and vanishes in a neighborhood of Ω_{4a} (cf. B.), since then the general case can be treated by an obvious limiting process. Furthermore, using a partition of unity, we may assume that ν has arbitrarily small compact support in $S - \Omega_{4a}$. More precisely, we assume that we are given a and b, $0 < b < a$ and a point $p_0 \in S$ such that we may form the domain $\Delta(p_0, 3b, 3a)$ described in B., and that every point of the support of ν has a ϕ-distance from p_0 not exceeding b. Introducing the natural parameter $\zeta(p) = \xi + i\eta$ belonging to ϕdz^2 at

p_0 and setting $\nu(\zeta(p)) = n(\zeta)$, we see that the relation to be proved becomes

$$\int_{-\infty}^{+\infty}\!\!\int \left\{\int_{-a}^{a} n(\xi+\tau,\eta)d\tau\right\} d\xi d\eta = 2a \int_{-\infty}^{+\infty}\!\!\int n(\xi,\eta)d\xi d\eta.$$

Its validity is, therefore, obvious.

12. Uniqueness proof

We are now in a position to prove Theorem I.

A. *Let S be a closed Riemann surface of genus $g > 1$. If $f : S \to S$ is a conformal homeomorphism homotopic to the identity, then f is the identity.*

This classical statement follows at once from the considerations of §6. Set $S = U/G$ and let $w : U \to U$ be the homeomorphism inducing f. If f is conformal, w must be conformal, that is an element of H. If f is homotopic to the identity, w induces the identity isomorphism on G. Hence w commutes with every element of G, and, since G must contain two non-commuting non-Euclidean translations, w is the identity and so is f.

This argument proves Theorem I for the case $K[f_0] = 1$.

B. Now let $f_0 : S \to S'$ be a Teichmüller mapping between two closed Riemann surfaces of genus $g > 1$, defined by the regular quadratic differential $\phi(z)dz^2 \not\equiv 0$ on S and some constant. Let $\psi(z')dz'^2$ be the regular quadratic differential on S' constructed in §8,E. Finally let $f : S \to S'$ be a quasiconformal mapping homotopic to f_0. We set $K[f_0] = K_0$, $K[f] = K$ and show first that

$$K \geq K_0. \tag{12.1}$$

The mapping $h = f_0 f^{-1}$ is a quasiconformal homeomorphism of S' onto itself which is homotopic to the identity, so that by Lemma B of the preceding section and by Schwarz's inequality

$$\iint_{S'} dA_\psi \leq \iint_{S'} \lambda^2_{h;\psi,\psi} dA_\psi. \tag{12.2}$$

On the other hand, it follows from §8,E. and the definitions given in §11,D. that we have

$$\lambda_{f;\phi,\psi}(p) = K_0 \lambda_{h;\psi,\psi}(f(p)), \tag{12.3}$$

$$j_{f_0;\phi,\psi} = K_0. \tag{12.4}$$

Noting property (11.5) of the function j and using relations (12.3), (12.4) together with inequality (11.7) we obtain

$$\iint_{S'} \lambda^2_{h,\psi,\psi} dA_\psi = K_0^{-2} \iint_{S'} \lambda^2_{f,\phi,\psi} dA_\psi = K_0^{-1} \iint_{S} \lambda_{f,\phi,\psi} dA_\phi$$

$$\leq (K/K_0) \iint_{S} j_{f,\phi,\psi} dA_\phi = (K/K_0) \iint_{S'} dA_\psi. \tag{12.5}$$

This inequality, together with (12.2), implies (12.1).

C. Assume now that the equality sign holds in (12.1). Then it must hold also in (12.5), so that by (11.7) we must have

$$\lambda^2_{f,\phi,\psi} = K_0 j_{f,\phi,\psi}. \tag{12.6}$$

Let $p_0 \in S$ be a point such that $\phi dz^2 \neq 0$ at p_0 and $\psi dz'^2 \neq 0$ at $f(p_0)$. Let $\zeta = \xi + i\eta$ be the natural parameter belonging to ϕdz^2 at p_0, and $\zeta' = \xi' + i\eta'$ and $\zeta^0 = \xi^0 + i\eta^0$ the natural parameters belonging to $\psi dz'^2$ at $f(p_0)$ and at $f_0(p_0)$, respectively. By §8, D., or by the definition of a Teichmüller mapping, we see that the function $\zeta^0(\zeta)$ defined by the mapping f_0 satisfies the Beltrami equation

$$\zeta^0_{\bar{\zeta}} = k_0 \bar{\zeta}^0_{\bar{\zeta}}, \qquad \left(k_0 = \frac{K_0 - 1}{K_0 + 1}\right).$$

On the other hand, by (12.6) and by inequality (3.1′) we have that the function $\zeta'(\zeta_0)$ defined by the mapping f satisfies the relations

$$(\xi'_\xi)^2 + (\eta'_\xi)^2 = K_0(\xi'_\xi \eta'_\eta - \xi'_\eta \eta'_\xi),$$

$$(\xi'_\xi)^2 + (\xi'_\eta)^2 + (\eta'_\xi)^2 + (\eta'_\eta)^2 \leq [K_0 + (1/K_0)](\xi'_\xi \eta'_\eta - \xi'_\eta \eta'_\xi).$$

A simple computation shows that these two relations imply that

$$\zeta'_{\bar{\zeta}} = k_0 \bar{\zeta}'_{\bar{\zeta}}.$$

Noting §4,B. we conclude that the function $\zeta'(\zeta_0)$ defined by the mapping h is analytic.

Thus the mapping h is conformal at all points of S', except perhaps at zeros of $\psi dz'^2$. By the theorem on removable singularities it is conformal also at these points. Since $h : S' \to S'$ is homotopic to the identity, it is the identity (cf. A.). Thus $f = f_0$ and Theorem I is proved.

13. Moduli of marked Riemann surfaces

In this section we prepare for the existence proof by associating with each marked Riemann surface of genus $g > 1$ a set of $6g-6$ real numbers.

A. Let $g > 1$ be a given integer. An ordered set $(A_1, \dots, A_{2g})$ of non-Euclidean translations will be called normalized if the following conditions are satisfied. (i) The group G generated by $(A_1, \dots, A_{2g})$ is a Fuchsian group with a compact fundamental region. (ii) The transformations $A_1, \dots, A_{2g}$ form a set of generators with the single defining relation

$$\prod_{j=1}^{g} (A_{2j}A_{2j-1}^{-1}A_{2j}^{-1}A_{2j-1}) = 1. \tag{13.1}$$

(Relation (13.1) holds also between matrices representing the A_j (cf. Fricke [6]), as can be seen by a continuity argument based on §7.D.) (iii) The repelling and attracting fixed points of A_{2g} are 0 and ∞, respectively, and the fixed points of A_{2g-1} are two real numbers the product of which has absolute value 1.

We shall denote the repelling and attracting fixed points and the invariants of A_j by σ_j, τ_j, and λ_j, respectively. Condition (iii) means therefore that

$$\sigma_{2g} = 0, \quad \tau_{2g} = \infty, \quad |\sigma_{2g-1}\tau_{2g-1}| = 1. \tag{13.2}$$

Actually we will always have $\sigma_{2g-1}\tau_{2g-1} = -1$, but this is of no importance for what follows.

The numbers $\sigma_1, \tau_1, \lambda_1, \sigma_2, \dots, \lambda_{2g-2}$ will be called the coordinates of the normalized set $(A_1, \dots, A_{2g})$.

B. *A normalized set of non-Euclidean translations is uniquely determined by its coordinates.*

PROOF. The coordinates determine the transformations $A_1, \dots, A_{2g-2}$ and hence, by (13.1), also the commutator $A_{2g}A_{2g-1}^{-1}A_{2g}^{-1}A_{2g-1} = B$. In view of (13.2) we may write

$$A_{2g-1}(z) = \frac{\alpha z + \beta}{\varepsilon\beta z + \delta}, \quad A_{2g}(z) = \lambda z$$

where

$$\lambda > 1; \ \varepsilon = \pm 1; \ \alpha, \beta, \delta \text{ are real}, \ \alpha\delta - \varepsilon\beta^2 = 1, \ \beta > 0. \tag{13.3}$$

Thus

$$B(z) = \frac{(\lambda\alpha\delta - \lambda^2\varepsilon\beta^2)z + (\lambda - \lambda^2)\beta\delta}{(\lambda - 1)\varepsilon\alpha\beta z + \lambda\alpha\delta - \varepsilon\beta^2} \tag{13.4}$$

and an easy computation shows that knowing $B(z)$ we can find numbers $\lambda, \varepsilon, \alpha, \beta, \delta$ satisfying (13.3), (13.4) in at most one way.

C. Using the same notations as in A. set $S = U/G$. Then S is a closed Riemann surface of genus g. For every $z \in U$ the natural isomorphism $\Lambda_{G,z}$ (cf. §5,E.) maps the normalized set $(A_1, \dots, A_{2g})$ onto a standard set of generators of the fundamental group $\pi_1(S, \Pi_G(z))$ and all standard sets so obtained are equivalent (cf. §9,A.). Denote their equivalence class by α. The marked Riemann surface (S, α) will be said to be defined by the normalized set $(A_1, \ldots, A_{2g})$. The point with the coordinates $\sigma_1, \tau_1, \lambda_1, \sigma_2, \tau_2, \ldots, \lambda_{2g-2}$ in the $(6g-6)$-dimensional Euclidean space will be called the representative point of (S, α).

D. *Every marked Riemann surface* (S, α) *of genus* $g > 1$ *is defined by a uniquely determined normalized set of non-Euclidean translations* (and thus has a representative point).

To prove this, represent S as U/G. The representation is unique, except that we may replace G by CGC^{-1}, $C \in H$. For some $z \in U$ set $\hat{A}_j = \Lambda_{G,z}^{-1}(a_j)$, $j = 1, 2, \ldots, 2g$, where $a = (a_1, \ldots, a_{2g})$ is some standard set in α at $\Pi_G(z)$. Since the group G contains only the identity and non-Euclidean translations (cf. §5,D.) the $\hat{A}_j$ are non-Euclidean translations. For the same reason any two of them have four distinct fixed points, for otherwise their commutator would not have two distinct fixed points. It is clear that the set $(\hat{A}_1, \ldots, \hat{A}_{2g})$ has properties (i) and (ii). Denote the fixed points of $\hat{A}_j$ by $\hat{\sigma}_j, \hat{\tau}_j$. There exists an element $C \in H$ which satisfies the conditions $C(\hat{\sigma}_{2g}) = 0$, $C(\hat{\tau}_{2g}) = \infty$, $|C(\sigma_{2g-1})C(\tau_{2g-1})| = 1$. Set $A_j = C\hat{A}_jC^{-1}$. Then the set $(A_1, \ldots, A_{2g})$ has properties (i)–(iii) and defines (S, α). It is clear that $(A_1, \ldots, A_{2g})$ is uniquely determined by (S, α).

E. Now let $(A_1, \ldots, A_{2g})$ and $(A_1', \ldots, A_{2g}')$ be two normalized sets of non-Euclidean translations which generate the groups G and G' and the marked Riemann surfaces (S, α) and (S', α'), respectively. Also, let w be a homeomorphism of U onto itself which satisfies the condition

$$A_j' = wA_jw^{-1}, \qquad j = 1, 2, \ldots, 2g. \tag{13.5}$$

Then $wGw^{-1} = G'$, so that w is compatible with G (cf. §6,G.) and therefore induces a homeomorphism $f: S \to S'$. Noting §6,E. we conclude that $f: (S, \alpha) \to (S', \alpha')$. This remark will be used in the existence proof given below.

14. Existence proof

We are now in a position to prove Theorem II. The proof hinges on the Riemann–Roch theorem, or rather on a corollary concerning regular quadratic differentials.

A. Throughout this section we denote by (S^0, α^0) a fixed marked Riemann surface of genus $g > 1$. This marked surface is defined by a normalized set of non-Euclidean translations $(A_1^0, \ldots, A_{2g}^0)$. We denote by $\sigma_1^0, \tau_1^0, \ldots, \sigma_{2g}^0, \tau_{2g}^0$ the fixed points of these translations; they satisfy, of course, the normalization condition (13.2). G^0 denotes the group generated by $(A_1^0, \ldots, A_{2g}^0)$.

We denote by M the set of all Beltrami coefficients $\mu(z)$, $z \in U$, compatible with G_0 (cf. §7,C.). The set is normed by setting $\|\mu\|$ = true maximum of $|\mu(z)|$, and a convergence concept is defined by requiring $\mu_n \to \mu$ to mean that $\|\mu_n\| \leq k < 1, n = 1, 2, \ldots$, and $\mu_n(z) \to \mu(z)$ almost everywhere.

We denote by C the set of all representative points of marked Riemann surfaces of genus g which are oriented similarly to (S^0, α^0), and by C^* the set of the representative points of marked Riemann surfaces with opposite orientation. C and C^* are disjoint subsets of E_{6g-6} (the $(6g-6)$-dimensional Euclidean space); their union is the set of representative points of all marked Riemann surfaces of genus g.

If ξ is a point in a Euclidean space, $|\xi|$ denotes its distance from the origin.

B. For $\mu \in M$ let $w^\mu(z)$ denote the uniquely determined solution of the Beltrami equation $w_{\bar{z}} = \mu(z)w_z$ which maps U homeomorphically onto itself and satisfies the conditions

$$(14.1) \qquad w^\mu(0) = 0, \qquad w^\mu(\infty) = \infty, \qquad |w^\mu(\sigma^0_{2g-1})w^\mu(\tau^0_{2g-1})| = 1$$

(cf. §7,D.). Set $A_j^\mu = (w^\mu)A_j^0(w^\mu)^{-1}, j = 1, 2, \ldots, 2g$. It is seen at once that $(A_1^\mu, \ldots, A_{2g}^\mu)$ is a normalized set of non-Euclidean translations, for the fixed points of A_j^μ are $w^\mu(\sigma_j^0)$ and $w^\mu(\tau_j^0)$. This set defines a marked Riemann surface (S^μ, α^μ) with the representative point $x = \Phi(\mu)$. Thus (cf. §13,E.) w^μ induces a homeomorphism $f^\mu : S \to S'$ which maps (S^0, α^0) onto (S^μ, α^μ). Since w^μ is orientation preserving, so is f^μ. Hence $\Phi(\mu) \in C$ and we have defined a mapping $\Phi : M \to C$. We note that

$$(14.2) \qquad K[f^\mu] = \frac{1 + \|\mu\|}{1 - \|\mu\|}.$$

C. Lemma 1. *The mapping $\Phi : M \to C$ is sequentially continuous and onto.*

Proof. Assume that $\mu_n \to \mu$ in M. By §4,E. and, in particular, by the remark made at the end of that paragraph, we conclude that $w^{\mu_n} \to w^\mu$ in the topology of W^* (cf. §5,A.). This implies that the fixed points and invariants of A^{μ_n} converge to those of A^μ and proves the continuity of our mapping.

Now let x be a given point of C, $(A_1, \ldots, A_{2g})$ the corresponding normalized set of non-Euclidean translations and (S, α) the marked Riemann surface defined by it. By the definition of C the marked Riemann surfaces (S^0, α^0) and (S, α) are similarly oriented. Hence (cf. §9,D.) there exists a quasiconformal mapping $f: (S^0, \alpha^0) \to (S, \alpha)$. By §7,C. and D. the mapping f is induced by a homeomorphism $w : U \to U$. This homeomorphism is quasiconformal and must satisfy the relations $A_j = wA_j^0w^{-1}$,

$j = 1, 2, \ldots, 2g$. In particular, it must satisfy the normalization condition (14.1). Hence w is of the form w^{μ}, $\mu \in M$, and $x = \Phi(\mu)$.

D. *The regular quadratic differentials on a closed Riemann surface S of genus $g > 1$ form a real linear vector space of dimension* $6g-6$.

This is a classical corollary of the Riemann–Roch theorem. Indeed, let $\omega(z)dz$ be some holomorphic differential of type $(1, 0)$ on S (Abelian differential of the first kind). Every regular quadratic differential on S is of the form $H(\omega dz)^2$ where H is a meromorphic function which is a multiple of the divisor of $(\omega dz)^2$. The number of linearly independent functions satisfying this condition can be computed by the Riemann–Roch theorem and equals $6g-6$.

E. It follows from D. that there exist $6g-6$ linearly independent functions, $\phi_1(z), \ldots, \phi_{6g-6}(z)$, $z \in U$, which are holomorphic solutions of the functional equation

$$\phi(A_j^0(z)) = \frac{\phi(z)}{(dA_j^0(z)/dz)^2}, \qquad j = 1, 2, \ldots, 2g \tag{14.3}$$

(cf. §8,B.). Using these functions we define a mapping of the open unit ball B in E_{6g-6} into the space M by setting $\mu = \Psi(\xi)$ where $\xi = (\xi_1, \ldots, \xi_{6g-6})$ is a point in B and $\mu(z)$ the function $\mu(z) \equiv 0$ if $\xi = 0$ and the function

$$\mu(z) = \left(\sum_{j=1}^{6g-6} \xi_j^2\right)^{1/2} \left(\sum_{j=1}^{6g-6} \xi_j \overline{\phi_j(z)}\right) \Bigg/ \left|\sum_{j=1}^{6g-6} \xi_j \phi_j(z)\right| \tag{14.4}$$

if $\xi \neq 0$ (cf. §8,B.). We note that $\|\Psi(\xi)\| = |\xi|$.

LEMMA 2. *The mapping $\Psi : B \to M$ is sequentially continuous.*

The proof is trivial.

F. Set $\Omega = \Phi\Psi$. The mapping $\Omega : B \to C$ has the following significance. If $x = \Omega(\xi)$, then there exists a Teichmüller mapping f_0 of (S^0, α^0) onto the marked Riemann surface represented by the point x. For this mapping we have (cf. §7,B.)

$$K[f_0] = \frac{1 + |\xi|}{1 - |\xi|}. \tag{14.5}$$

Noting Lemmas 1 and 2, we obtain

LEMMA 3. *The mapping $\Omega : B \to C$ is continuous.*

LEMMA 4. *The mapping $\Omega : B \to C$ is one-to-one.*

PROOF. If $x = \Omega(\xi_1) = \Omega(\xi_2)$ and $\xi_1 \neq \xi_2$, then there exist two distinct Teichmüller mappings of (S^0, α^0) onto the marked Riemann surface represented by x. This is impossible by the corollary to Theorem I (cf. §10,A.).

LEMMA 5. *The mapping $\Omega : B \to C$ is a homeomorphism and $\Omega(B)$ is open.*

This follows from Lemmas 3 and 4 and Brouwer's theorem on invariance of domain.

G. Theorem II is contained in the following:

LEMMA 6. *The mapping* $\Omega : B \to C$ *is onto, so that* C *is homeomorphic to* E_{6g-6}.

In fact, this lemma implies that for any marked Riemann surface (S, α) of genus g which is similarly oriented to (S^0, α^0) there exists a Teichmüller mapping $f_0 : (S^0, \alpha^0) \to (S, \alpha)$.

Before proving this lemma we note:

LEMMA 7. *If* $\Omega(\xi) = \Phi(\mu)$, *then* $\|\mu\| \geq |\xi|$.

PROOF. Set $x = \Omega(\xi)$ and let (S, α) be the marked Riemann surface represented by this point. Then there exists a quasiconformal mapping $f^\mu : (S^0, \alpha^0) \to (S, \alpha)$ satisfying (14.2) and a Teichmüller mapping $f_0 : (S^0, \alpha^0) \to (S, \alpha)$ satisfying (14.5). The assertion follows from Theorem I.

H. Now we prove Lemma 6. Let x_1 be a given point in C. By Lemma 1 there is a $\mu \in M$ with $\Phi(\mu) = x_1$. For every $t, 0 \leq t \leq 1$, we have $t\mu \in M$. Set $x_t = \Phi(t\mu)$. The point x_t depends continuously on t, by virtue of Lemma 3.

Let θ denote the set of those values of t for which $x_t \in \Omega(B)$. This set is not empty since it contains the origin: x_0 is the representative point of (S^0, α^0). The set θ is open in view of Lemma 5. We want to show that it is closed, since this would imply that it contains 1, so that $x_1 \in \Omega(B)$.

Assume that there is a sequence $\{t_j\}$ in θ with $t_j \to t_\infty$. Then there exists a sequence of points $\{\xi_j\}$ in B with $x_{t_j} = \Omega(\xi_j)$. By Lemma 7 $|\xi_j| \leq t_j \|\mu\| \leq \|\mu\| < 1$. Hence we may assume, selecting if need be a subsequence, that $\xi_j \to \xi_\infty \in B$. By Lemma 3 we have that $\Omega(\xi_\infty) = \lim \Omega(\xi_j) = \lim x_{t_j} = x_{t_\infty}$. Hence, $t_\infty \in \theta$, q.e.d.†

I. The following remarks are a digression from our main subject. With the notations of A. and B. set $\hat{w}^\mu(z) = -\overline{w^\mu(z)}$. Then $\hat{w}^\mu : U \to U$ is an orientation-reversing homeomorphism and one sees easily that it induces a mapping $\hat{f}^\mu$ of (S^0, α^0) onto a marked Riemann surface $(\hat{S}^\mu, \hat{\alpha}^\mu)$ with an opposite orientation. Let $\sigma_j^\mu, \tau_j^\mu, \lambda_j^\mu$ and $\hat{\sigma}_j^\mu, \hat{\tau}_j^\mu, \hat{\lambda}_j^\mu$ be the fixed points and invariants of the normalized sets $(A_1^\mu, \ldots, A_{2g}^\mu)$ and $(\hat{A}_1^\mu, \ldots, \hat{A}_{2g}^\mu)$ defining the marked Riemann surfaces (S^μ, α^μ) and $(\hat{S}^\mu, \hat{\alpha}^\mu)$, respectively. Since w^μ is orientation preserving and satisfies (14.1), $\sigma_{2g-1}^\mu \sigma_{2g-1}^0 > 0$. On the other hand, we have that $\hat{\sigma}_j^\mu = -\sigma_j^\mu$, $\hat{\tau}_j^\mu = -\tau_j^\mu$, $\hat{\lambda}_j^\mu = \lambda_j^\mu$. Thus we obtain the following conclusions.

(a) Let $(A_1, \ldots, A_{2g})$ and $(B_1, \ldots, B_{2g})$ be normalized sets of

† The reader will notice that we do not need the fact that the dimension δ of the space of regular quadratic differentials on S is $6g-6$ but only the inequality $\delta \geq 6g-6$. On the other hand, Ahlfors needs for his variational proof the fact that $\delta < \infty$.

Non-Euclidean translations. The marked Riemann surfaces defined by these sets are similarly oriented if and only if $\sigma(A_{2g-1})\sigma(B_{2g-1}) > 0$.

(b) The set C^* is obtained from the set C by changing the sign of $4g-4$ coordinates in E_{6g-6}.

15. The space of marked Riemann surfaces

The preceding considerations contain, implicitly, the proof of Theorem III. For the sake of completeness, we spell out the details.

A. Let C have the same meaning as in the preceding section. The points of C are thus representative points of marked Riemann surfaces of fixed genus $g > 1$, all with the same orientation. The Teichmüller distance, $d(x_1, x_2)$, between two points of C is, by definition, the Teichmüller distance between the corresponding marked Riemann surfaces (cf. §10, C.). By Lemmas 1, 5, and 6 of the preceding section, we know that the mapping Ω establishes a homeomorphism between the unit ball B in E_{6g-6} and the set $C \subset E_{6g-6}$, with respect to the Euclidean distance function $|x_1 - x_2|$. Hence Theorem III will be proved once we establish the topological equivalence of the Euclidean and Teichmüller metrics in C.

B. Let x_n, $n = 0, 1, 2, \ldots$, be points on C. Without loss of generality we may assume that x_0 is the representative point of the marked Riemann surface (S^0, α^0) considered in §14. We set $\xi_n = \Omega^{-1}(x_n)$ and note that $\xi_0 = 0$.

We must show that each of the two relations: $d(x_0, x_n) \to 0, |x_0 - x_n| \to 0$, implies the other. Set, using the notations of the preceding section, $\mu_n = \Psi(\xi_n)$. Then the Teichmüller mapping $f_{0,n}$ of (S^0, α^0) onto the marked Riemann surface represented by x_n has as its Beltrami differential μ_n. Hence (cf. §14, E.) $K[f_{0,n}] = (1 + |\xi_n|)/(1 - |\xi_n|)$ so that (cf. §10, C.) $d(x_0, x_n) = \log[(1 + |\xi_n|)/(1 - |\xi_n|)]$. Thus $d(x_0, x_n) \to 0$ if and only if $\xi_n \to 0$ and, since Ω is a homeomorphism, if and only if $|x_0 - x_n| \to 0$.

Note added in proof. The space R_g of marked Riemann surfaces has a natural complex analytic structure as was proved by Ahlfors (see his paper in this volume). Other proofs were given by Spencer and Kodaira, Weil, and the author. For a report on these and related matters see the author's paper *Spaces of Riemann Surfaces* in the Proceedings of the Edinburgh Congress.

References

[1] L. V. Ahlfors, On quasi-conformal mappings, *Journal d'Analyse Mathematique 4* (1954), pp. 1–58.

[2] L. V. Ahlfors and Lipman Bers, To appear.

[3] Lipman Bers, On a theorem of Mori and the definition of quasiconformality, *Transactions of the American Mathematical Society 84* (1957), pp. 78–84.

[4] LIPMAN BERS and LOUIS NIRENBERG, On a representation theorem for linear elliptic systems with discontinuous coefficients and its applications, *Convegno Internazionale sulle Equazioni Derivate e Parziali* (1954), pp. 111–140.

[5] B. V. BOYARSKII, Homeomorphic solutions of Beltrami systems, *Doklady 102* (1955), pp. 661–664. (Russian).

[6] ROBERT FRICKE and FELIX KLEIN, *Vorlesungen über die Theorie der automorphen Funktionen 1*, Leipzig (1926).

[7] K. O. FRIEDRICHS, The identity of weak and strong extensions of differential operators, *Transactions of the American Mathematical Society 55* (1944), pp. 132–151.

[8] HERBERT GRÖTZSCH, Über die Verzerrung bie schlichten nichtkonformen Abbildungen und über eine damit zusammenhängende Erweiterung des Picardschen Satzes, *Leipz. Ber. 80* (1928), pp. 503–507.

[9] ———, Über möglichst konforme Abbildungen von schlichten Bereichen, *Leipz. Ber. 84* (1932) pp. 114–120.

[10] J. A. JENKINS, A new criterion for quasi-conformal mappings, *Annals of Mathematics 65* (1957), pp. 208–214.

[11] M. A. LAVRENT'EV, Sur une classe de représentation continues, *Mat. Sbornik 42* (1935), pp. 407–423.

[12] ———, A fundamental theorem of the theory of quasi-conformal mapping of plane regions, *Izvestya Akademii Nauk S.S.S.R. 12* (1948), pp. 513–554. (Russian).

[13] W. MANGLER, Die Klassen von topologischen Abbildungen einer geschlossenen Flache auf sich, *Mathematische Zeitschrift 44* (1939), pp. 541–554.

[14] AKIRA MORI, On quasi-conformality and pseudoanalyticity, *Transactions of the American Mathematical Society 84* (1957), pp. 56–57.

[15] ———, On an absolute constant in the theory of quasi-conformal mappings, *Journal of the Mathematical Society of Japan 8* (1956), pp. 156–166.

[16] C. B. MORREY, On the solutions of quasilinear elliptic partial differential equations, *Transactions of the American Mathematical Society 43* (1938), pp. 126–166.

[17] I. N. PESIN, Metric properties of quasi-conformal mappings, *Mat. Sbornik 40* No. 82 (1956), pp. 281–294.

[18] ALBERT PFLUGER, Quasikonforme Abbildungen und logarithmische Kapazität, *Annales de l'Institut Fourier 2* (1950), pp. 69–80.

[19] PLUTARCH, *Lives*, Loeb's Classical Library 3, p. 5.

[20] V. I. SMIRNOV, *Higher Mathematics 5*, Leningrad (1949–51). (Russian).

[21] S. L. SOBOLEV, *Some applications of functional analysis to mathematical physics*, Leningrad 1950. (Russian).

[22] OSWALD TEICHMÜLLER, Extremale quasikonforme Abbildungen und quadratische differentiale, *Preuss. Akad. 22* (1940).

[23] ———, Bestimmung der extremalen quasikonformen Abbildungen bei geschlossenen orientierten Riemannschen Flächen, *Preuss. Akad. 4* (1943).

[24] L. I. VOLKOVISKII, *Quasi-conformal mappings*, Lwōw University (1954). (Russian).

On Compact Analytic Surfaces

Kunihiko Kodaira
PRINCETON UNIVERSITY AND
INSTITUTE FOR ADVANCED STUDY

THE present note is a preliminary report on a study of structures of compact analytic surfaces.

1. Let V be a compact analytic surface, i.e. a compact complex manifold of complex dimension 2. Let $\mathscr{M}(V)$ be the field of all meromorphic functions on V and denote by dim $\mathscr{M}(V)$ the degree of transcendency of $\mathscr{M}(V)$ over the field **C** of all complex numbers. By a result due to Chow [2] and Siegel [9], we have

$$\dim \mathscr{M}(V) \leqq 2.$$

We denote by $p_g(V)$ the *geometric genus* of V, by c_1 the first Chern class of V and by $c_1^2(V)$ the value of $c_1^2 = c_1 \cdot c_1$ on the fundamental 4-cycle V. The geometric genus $p_g(V)$ is, by definition, the number of linearly independent holomorphic 2-forms on V. By the *canonical bundle* K of V we mean the complex line bundle over V of forms of type (2, 0). We note that the characteristic class $c(K)$ of K is equal to $-c_1$.

THEOREM 1. (Chow and Kodaira [3], Kodaira [6].) *A compact analytic surface V is a non-singular algebraic surface imbedded in a projective space if and only if* $\dim \mathscr{M}(V) = 2$.

Let Δ be a non-singular algebraic curve (i.e. a compact Riemann surface). We say that V is an *analytic fibre space of curves of genus π over Δ* if there exists a holomorphic map Φ of V *onto* Δ such that the inverse image $C_u = \Phi^{-1}(u)$ of any "general point" u of Δ is a non-singular irreducible curve of genus π, and we call Φ the canonical projection of V onto Δ, the inverse image $\Phi^{-1}(u)$ of each point $u \in \Delta$ a *fibre* of V. In what follows we are mainly concerned with analytic fibre spaces of *elliptic* curves i.e. curves of genus 1.

THEOREM 2. (Kodaira [6].) *If* $\dim \mathscr{M}(V) = 1$, *then V is an analytic fibre space of elliptic curves over a curve Δ and the canonical projection $\Phi : V \rightarrow \Delta$ induces an isomorphism*

$$\Phi^* : \mathscr{M}(V) \cong \mathscr{M}(\Delta),$$

where $\mathscr{M}(\Delta)$ denotes the field of meromorphic functions on Δ. Moreover V contains no irreducible curve other than the components of the fibres of V.

We denote by (CD) the *intersection multiplicity* of two curves C and D on V and write (C^2) for (CC). A curve C on V will be called an *exceptional curve* (of the the first kind) if C is a non-singular rational curve with $(C^2) = -1$ (compare Zariski [10], pp. 36–41. By an exceptional curve we mean always an exceptional curve of the first kind). For an arbitrary point $p \in V$ we denote by Q_p the *quadratic transformation* with the center p (see Hopf [5]). Moreover we call any surface $W = \ldots Q_{p_3} Q_{p_2} Q_{p_1}(V)$ obtained from V by applying a finite number of quadratic transformations a *quadratic transform* of V. We recall that $S = Q_p(p)$ is an exceptional curve on the quadratic transform $\hat{V} = Q_p(V)$ and that Q_p^{-1} is a holomorphic map of $\hat{V}$ onto V which is biregular between $\hat{V} - S$ and $V - p$. Moreover Q_p^{-1} induces an isomorphism: $\mathcal{M}(Q_p(V)) \cong \mathcal{M}(V)$.

THEOREM 3. *Assume that* dim $\mathcal{M}(V) \geqq 1$. *If V contains an exceptional curve S, then there exists a compact analytic surface W and a point $p \in W$ such that $V = Q_p(W)$ and $S = Q_p(p)$.*

This theorem has been established for algebraic surfaces by Castelnuovo and Enriques [1] (compare also Kodaira [7]). In view of Theorem 1, it suffices therefore to consider the case in which dim $\mathcal{M}(V) = 1$. Now, if dim $\mathcal{M}(V) = 1$, V is an analytic fibre space of elliptic curves over a curve Δ and the canonical projection Φ maps S onto a single point u on Δ. Let U be a small neighborhood of u on Δ. A detailed analysis of the structure of the fibre space V shows that the neighborhood $\Phi^{-1}(U)$ of S can be imbedded in an algebraic surface. Hence the theorem is reduced to the case of algebraic surfaces.

As to compact analytic surfaces with no meromorphic functions except constants, we have the following

THEOREM 4. (Kodaira [6].) *Let V be a compact Kähler surface with* dim $\mathcal{M}(V) = 0$. *The geometric genus $p_g(V)$ of V is equal to* 1 *and the first Betti number $b_1(V)$ of V is equal to either* 4 *or* 0. *If $b_1(V) = 4$, then V is a quadratic transform of a complex torus* (*of complex dimension* 2). *If $b_1(V) = 0$ and if V contains no exceptional curve, then the canonical bundle K of V is trivial and the second Betti number $b_2(V)$ of V is equal to* 22.

It can be shown that $c_1^2(V) \leqq 0$ if V is an analytic fibre space of elliptic curves over a curve, while, with the help of Riemann-Roch's theorem, we infer readily that dim $\mathcal{M}(V) \geqq 1$ if $p_g(V) \geqq 1$ and if $c_1^2(V) > 0$. Hence we obtain from Theorems 1, 2, and 4 the following

THEOREM 5. *If V is a compact Kähler surface and if $c_1^2(V) > 0$, then V is an algebraic surface.*

2. Let V be an analytic fibre space of elliptic curves over a curve Δ having no exceptional curve and let Φ be the canonical projection of V onto Δ. For an arbitrary point $a \in \Delta$, we denote by τ_a the local uniformization variable on Δ with the center a and by $\tau_a(u)$ the value of τ_a at a point u in a neighborhood of a on Δ. Moreover we denote by (z_1, z_2) a system of local coordinates on V. Now, if

$$\left|\frac{\partial \tau_a(\Phi(z))}{\partial z_1}\right|^2 + \left|\frac{\partial \tau_a(\Phi(z))}{\partial z_2}\right|^2 > 0 \tag{1}$$

at each point z on $\Phi^{-1}(u)$, we call $C_u = \Phi^{-1}(u)$ a *regular fibre* of V. Clearly each regular fibre C_u is a non-singular elliptic curve. By a theorem of Bertini, there exists a finite set $\{a_\rho\}$ of points a_ρ, $\rho = 1, 2, 3, \ldots$, on Δ such that $C_u = \Phi^{-1}(u)$ is a regular fibre for each point $u \in \Delta - \{a_\rho\}$. In this section we assume that the fibres $\Phi^{-1}(a_\rho)$ are *not* regular. Clearly $\tau_{a_\rho}(\Phi) : z \to \tau_{a_\rho}(\Phi(z))$ is a holomorphic function defined on a neighborhood of $\Phi^{-1}(a_\rho)$ which vanishes at each point on $\Phi^{-1}(a_\rho)$. By the *singular fibre* C_{a_ρ} of V over a_ρ we mean the divisor of the holomorphic function $\tau_{a_\rho}(\Phi)$. We write each singular fibre C_{a_ρ} in the form

$$C_{a_\rho} = \sum_s n_{\rho s} \Theta_{\rho s},$$

where $\Theta_{\rho s}$ are irreducible curves and $n_{\rho s}$ are positive integers. Clearly $\Phi^{-1}(a_\rho)$ coincides with the union of the curves $\Theta_{\rho s}$; thus C_{a_ρ} may be considered as the inverse image of a_ρ attached with the proper multiplicities $n_{\rho s}$.

The expression $\sum_s n_{\rho s}\Theta_{\rho s}$ may be regarded as a 2-*cycle* on the polyhedron $\Phi^{-1}(a_\rho) = \bigcup_s \Theta_{\rho s}$. By the *type* of the singular fibre $C_{a_\rho} = \sum_s n_{\rho s}\Theta_{\rho s}$ we mean the topological structure of the polyhedron $\Phi^{-1}(a_\rho)$ together with the homology class of the 2-cycle $\sum n_{\rho s}\Theta_{\rho s}$ on $\Phi^{-1}(a_\rho)$. It is not difficult to determine all possible types of singular fibres of V. We denote these types by the symbols ${}_m\mathrm{I}_b$, I_b^*, II, II*, III, III*, IV, IV*. In order to describe the characteristic properties of each type (see Figure 1), we write $\sum n_s \Theta_s$ for $\sum n_{\rho s}\Theta_{\rho s}$ and denote by $\Theta_s \cdot \Theta_t$ the *intersection cycle* of Θ_s and Θ_t. For instance the formula $\Theta_s \cdot \Theta_t = 2p_1 + 3p_2$ indicates that Θ_s and Θ_t meet at two points p_1 and p_2 with intersection multiplicities 2 and 3, respectively.

${}_m\mathrm{I}_0$: $C_{a_\rho} = m\Theta_0$, $m = 2, 3, 4, \ldots$, where Θ_0 is a non-singular elliptic curve.

${}_m\mathrm{I}_1$: $C_{a_\rho} = m\Theta_0$, $m = 1, 2, 3, \ldots$, where Θ_0 is a rational curve with one ordinary double point.

${}_m\mathrm{I}_2$: $C_{a_\rho} = m\Theta_0 + m\Theta_1$, $m = 1, 2, 3, \ldots$, where Θ_0 and Θ_1 are non-singular rational curves with $\Theta_0 \cdot \Theta_1 = p_1 + p_2$.

II : $C_{a_\rho} = \Theta_0$ is a rational curve with one cusp.

III : $C_{a_\rho} = \Theta_0 + \Theta_1$, where Θ_0 and Θ_1 are non-singular rational curves with $\Theta_0 \cdot \Theta_1 = 2p$.

IV : $C_{a_\rho} = \Theta_0 + \Theta_1 + \Theta_2$, where Θ_0, Θ_1, Θ_2 are non-singular rational curves and $\Theta_0 \cdot \Theta_1 = \Theta_1 \cdot \Theta_2 = \Theta_2 \cdot \Theta_0 = p$.

The singular fibres of the types ${}_m\mathrm{I}_b$, $b \geq 3$, I_b^*, II*, III*, IV* are composed of non-singular rational curves $\Theta_0, \Theta_1, \ldots, \Theta_s, \ldots$ in such a way that $(\Theta_s \Theta_t) \leq 1$ (i.e. Θ_s and Θ_t have at most one simple intersection point) for $s < t$

and $\Theta_r \cap \Theta_s \cap \Theta_t$ is empty for $r < s < t$. These types are therefore described completely by showing *all pairs* Θ_s, Θ_t *with* $(\Theta_s\Theta_t) = 1$ besides $\sum n_s\Theta_s$.

${}_m\mathrm{I}_b$: $C_{a_\rho} = m\Theta_0 + m\Theta_1 + \ldots + m\Theta_{b-1}, m = 1, 2, 3, \ldots, b = 3, 4, 5, \ldots,$ $(\Theta_0\Theta_1) = (\Theta_1\Theta_2) = \ldots = (\Theta_s\Theta_{s+1}) = \ldots = (\Theta_{b-2}\Theta_{b-1}) = (\Theta_{b-1}\Theta_0) = 1.$

I_b^* : $C_{a_\rho} = \Theta_0 + \Theta_1 + \Theta_2 + \Theta_3 + 2\Theta_4 + 2\Theta_5 + \ldots + 2\Theta_{4+b}, b = 0, 1, 2, \ldots, (\Theta_0\Theta_4) = (\Theta_1\Theta_4) = (\Theta_2\Theta_{4+b}) = (\Theta_3\Theta_{4+b}) = (\Theta_4\Theta_5) = (\Theta_5\Theta_6) = \ldots = (\Theta_{3+b}\Theta_{4+b}) = 1.$

II* : $C_{a_\rho} = \Theta_0 + 2\Theta_1 + 3\Theta_2 + 4\Theta_3 + 5\Theta_4 + 2\Theta_5 + 4\Theta_6 + 3\Theta_7 + 6\Theta_8,$ $(\Theta_0\Theta_1) = (\Theta_1\Theta_2) = (\Theta_2\Theta_3) = (\Theta_3\Theta_4) = (\Theta_5\Theta_6) = (\Theta_4\Theta_8) = (\Theta_6\,\Theta_8) = (\Theta_7\Theta_8) = 1.$

III* : $C_{a_\rho} = \Theta_0 + 2\Theta_1 + 3\Theta_2 + \Theta_3 + 2\Theta_4 + 3\Theta_5 + 2\Theta_6 + 4\Theta_7, (\Theta_0\Theta_1) = (\Theta_1\Theta_2) = (\Theta_3\Theta_4) = (\Theta_4\Theta_5) = (\Theta_2\Theta_7) = (\Theta_5\Theta_7) = (\Theta_6\Theta_7) = 1.$

IV* : $C_{a_\rho} = \Theta_0 + 2\Theta_1 + \Theta_2 + 2\Theta_3 + \Theta_4 + 2\Theta_5 + 3\Theta_6, (\Theta_0\Theta_1) = (\Theta_2\Theta_3) = (\Theta_4\Theta_5) = (\Theta_1\Theta_6) = (\Theta_3\Theta_6) = (\Theta_5\Theta_6) = 1.$

A singular fibre $C_{a_\rho} = \sum_s n_{\rho s}\Theta_{\rho s}$ will be called *simple* or *multiple* according as min $\{n_{\rho s}\} = 1$ or $\geqq 2$. It is clear that C_{a_ρ} is a multiple singular fibre if and only if C_{a_ρ} is of type ${}_m\mathrm{I}_b$, $m \geqq 2$. Suppose that C_{a_ρ}, $1 \leqq \rho \leqq l$, are of types ${}_{m_\rho}\mathrm{I}_{b_\rho}$, $m_\rho \geqq 2$, respectively, and that C_{a_ρ}, $\rho \geqq l + 1$, are simple. Let m_0 be the l.c.m. of $m_1, \ldots, m_\rho, \ldots, m_l$, and let $d = m_1m_2 \ldots m_l$. Moreover let a_0 be an arbitrary point on $\Delta - \{a_\rho\}$. Then there exists a d-fold abelian covering surface $\tilde{\Delta}$ of Δ which is unramified over $\Delta - \{a_0, a_1, \ldots, a_l\}$ and has d/m_ρ branch points $b_{\rho k}$, $k = 1, 2, \ldots, d/m_\rho$ of order $m_\rho - 1$ over each point a_ρ, $\rho = 0, 1, 2, \ldots, l$. Letting ϖ be the canonical projection of $\tilde{\Delta}$ onto Δ, we define the analytic fibre space $\tilde{V}$ over $\tilde{\Delta}$ *induced from* V *by the map* ϖ to be the *minimal non-singular model* of the subvariety of $V \times \tilde{\Delta}$ consisting of all points $(z, \tilde{u}) \in V \times \tilde{\Delta}$ satisfying $\Phi(z) = \varpi(\tilde{u})$. The canonical projection $\tilde{\Phi} : \tilde{V} \to \tilde{\Delta}$ is induced from the projection $V \times \tilde{\Delta} \to \tilde{\Delta}$. $\tilde{V}$ is also an analytic fibre space of elliptic curves. It is clear that, for $\tilde{u} \neq b_{\rho k}$, the fibre $\tilde{C}_{\tilde{u}}$ of $\tilde{V}$ over $\tilde{u}$ has the same type as the fibre C_u of V over $u = \varpi(\tilde{u})$ and, in particular, $\tilde{C}_{\tilde{u}}$ is regular if $\varpi(\tilde{u}) \notin \{a_\rho\}$, while the fibre $\tilde{C}_{b_{\rho k}}$, $1 \leqq \rho \leqq l$, is of type ${}_1\mathrm{I}_{m_\rho b_\rho}$ and $\tilde{C}_{b_{0k}}$ is regular. Thus the induced fibre space $\tilde{V}$ is free from multiple singular fibres. On the other hand $\tilde{V}$ is a d-fold abelian covering manifold of V which is unramified over $V - C_{a_0}$ and has d/m_0 branch curves $\tilde{C}_{b_{0k}}$, $k = 1, 2, \ldots, d/m_k$, of order $m_0 - 1$ over C_{a_0}. The covering map $\Pi : \tilde{V} \to V$ is induced from the projection $V \times \tilde{\Delta} \to V$. Thus we obtain the following

THEOREM 6. *An analytic fibre space* V *of elliptic curves over a curve* Δ *having no exceptional curve induces over a suitably chosen finite ramified* *abelian covering surface $\tilde{\Delta}$ of Δ an analytic fibre space $\tilde{V}$ of elliptic curves free from multiple singular fibres. $\tilde{V}$ is a finite abelian covering of V having branch curves over a regular fibre C_{a_0} of V and contains no exceptional curve. V is represented as the factor space: $V = \tilde{V}/\mathfrak{G}$, where $\mathfrak{G}$ is the covering transformation group of $\tilde{V}$ over V.*

3. Now let V be an analytic fibre space of elliptic curves over Δ having neither exceptional curve nor multiple singular fibre and let $\Phi : V \to \Delta$ be

FIGURE 1

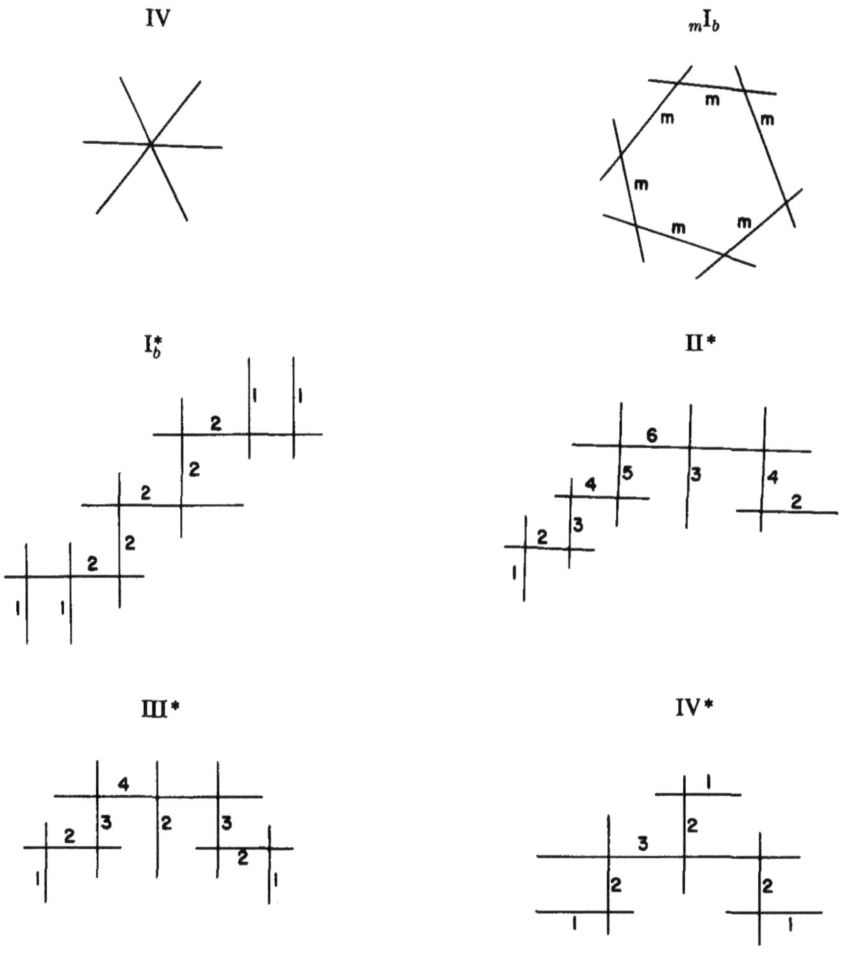

Each line represents Θ_{ps}; the integers attached to the line gives n_{ps}.

the canonical projection. Moreover let $\{a_\rho\}$ be a finite set of points $a_1, \ldots, a_\rho, \ldots, a_r$ on Δ such that the fibre $C_u = \Phi^{-1}(u)$ is regular for $u \neq a_\rho$ and let $\Delta' = \Delta - \{a_\rho\}$, where, for convenience, we do *not* assume that each fibre C_{a_ρ} is singular. C_{a_ρ} is therefore either a regular fibre or a singular fibre of one of the types I_b, I_b^*, II, II*, III, III*, IV, IV*, where we write I_b for ${}_1I_b$.

The restriction $V \mid \Delta' = \Phi^{-1}(\Delta') = \bigcup_{u\in\Delta'} C_u$ of V to Δ' is a differentiable fibre bundle of tori over Δ' and therefore the first homology groups $H_1(C_u, \mathbf{Z})(\cong \mathbf{Z} \oplus \mathbf{Z})$ of the fibres C_u with coefficients in the integers $\mathbf{Z}$ forms a locally constant sheaf

$$G' = \bigcup_{u\in\Delta'} H_1(C_u, \mathbf{Z})$$

over Δ' in a canonical manner. Let E_ρ be a circular neighborhood of a_ρ on Δ and let $E'_\rho = E_\rho - a_\rho$. Then the group $\Gamma(G' \mid E'_\rho)$ of sections of G' over E'_ρ is independent of the size of E'_ρ. We extend G' to a sheaf G over Δ by defining $\Gamma(G' \mid E'_\rho)$ to be the stalk G_{a_ρ} of G over a_ρ:

$$G = \bigcup_\rho G_{a_\rho} \cup G', \quad G_{a_\rho} = \Gamma(G' \mid E'_\rho)$$

and we call G the *homological invariant* of the fibre space V over Δ.

We denote by $J(\omega)$ the *elliptic modular function* defined on the upper half plane $\mathbf{C}^+ = \{\omega \mid \Im\omega > 0\}$. As is well known, $\omega \rightarrow J(\omega)$ is a holomorphic map of $\mathbf{C}^+$ onto the whole plane $\mathbf{C}$ and the equality $J(\omega) = J(\omega')$ holds if and only if ω' is obtained from ω by a *modular transformation*

$$S : \omega \rightarrow \omega' = \frac{a\omega + b}{c\omega + d}, \qquad ad - bc = 1,$$

where a, b, c, d are rational integers. For each point $u \in \Delta'$, we represent the elliptic curve C_u as a complex torus with periods $(\omega(u), 1)$, $\Im\omega(u) > 0$, and set

$$\mathcal{J}(u) = J(\omega(u)).$$

$\omega(u)$ is determined by C_u uniquely up to a modular transformation and *depends holomorphically* on $u \in \Delta'$. Hence $\mathcal{J}(u)$ is a single-valued holomorphic function of u defined on $\Delta' = \Delta - \{a_\rho\}$. Moreover it can be shown that each point a_ρ is either a pole or a removable singular point of $\mathcal{J}(u)$. Thus *$\mathcal{J}(u)$ is a meromorphic function on Δ*. We call $\mathcal{J}(u)$ the *functional invariant* of the fibre space V of elliptic curves over Δ. By a suitable choice of the finite set $\{a_\rho\}$, we may assume that $\mathcal{J}(u) \neq 0, 1, \infty$ for $u \in \Delta' = \Delta - \{a_\rho\}$.

Suppose conversely that an algebraic curve Δ of genus p and a meromorphic function $\mathcal{J}(u)$ on Δ are given. Take a finite set $\{a_\rho\}$ of points on Δ such that $\mathcal{J}(u) \neq 0, 1, \infty$ for $u \in \Delta' = \Delta - \{a_\rho\}$. Then there exists one and only one multi-valued holomorphic function $\omega(u)$ with

$\Im\omega(u) > 0$ on Δ' satisfying $J(\omega(u)) = \mathscr{J}(u)$. Take a point o on Δ' and suppose that each element of the fundamental group $\pi_1(\Delta')$ of Δ' is represented by a closed arc β on Δ' starting and ending at o. By the analytic continuation along the arc β, $\omega(u)$ is transformed into $S_\beta\omega(u)$, where S_β is a modular transformation depending only on the homotopy class of β relative to o. We indicate this fact by the formula

$$\omega(u) \xrightarrow{\beta} S_\beta\omega(u) = \frac{a_\beta\omega(u) + b_\beta}{c_\beta\omega(u) + d_\beta}. \tag{2}$$

The correspondence $\beta \to S_\beta$ gives a representation of $\pi_1(\Delta')$, provided that we take a fixed branch of $\omega(u)$ at the starting point o of β. In the particular case in which $\omega(u) = \omega_0$ is a constant, the condition (2), which states simply that ω_0 is a fixed point of S_β, may not be sufficient to determine the modular transformation S_β. Then we choose S_β arbitrarily, provided that S_β satisfies (2) and that $\beta \to S_\beta$ gives a representation of $\pi_1(\Delta')$. For each modular transformation S_β, $\beta \in \pi_1(\Delta')$, we choose

$$(S_\beta) = \pm\begin{pmatrix} a_\beta & b_\beta \\ c_\beta & d_\beta \end{pmatrix}$$

in such a way that $\beta \to (S_\beta)$ gives a representation: $\pi_1(\Delta') \to SL(2, \mathbf{Z})$. This is possible in $2p + r - 1$ different manners, provided that $r \geqq 1$, since $\pi_1(\Delta')$ is generated by $2p + r$ generators $\beta_1, \beta_2, \ldots, \beta_{2p}, \alpha_1, \ldots, \alpha_r$ with the single relation:

$$\beta_1\beta_2\beta_1^{-1}\beta_2^{-1}\beta_3\beta_4 \ldots \beta_{2p-1}^{-1}\beta_{2p}^{-1}\alpha_1\alpha_2 \ldots \alpha_r = 1.$$

The representation $\beta \to (S_\beta)$ defines a locally constant sheaf G' over Δ' whose stalks are isomorphic to $\mathbf{Z} \oplus \mathbf{Z}$ and G' can be extended uniquely to the sheaf $G = \bigcup_\rho G_{a_\rho} \cup G'$ over Δ, where $G_{a_\rho} = \Gamma(G' \mid E'_\rho)$. Under these circumstances we say that *the sheaf G belongs to the meromorphic function $\mathscr{J}(u)$*. It can be shown that, given an analytic fibre space V of elliptic curves over Δ free from multiple singular fibres, *the homological invariant G of V belongs to the functional invariant $\mathscr{J}(u)$ of V.*

DEFINITION 1. *Given an algebraic curve Δ, a meromorphic function $\mathscr{J} = \mathscr{J}(u)$ on Δ and a sheaf G over Δ belonging to $\mathscr{J}$, we denote by $\mathscr{F}(\mathscr{J}, G)$ the family of all fibre space V of elliptic curves over Δ free from multiple singular fibres and from exceptional curves having the given invariants $\mathscr{J}$ and G.*

We say that the analytic fibre spaces V_1 and V_2 of (elliptic) curves over Δ are *analytically equivalent* if and only if there exists an analytic homeomorphism $f: V_1 \to V_2$ such that $\Phi_2 f = \Phi_1$, where Φ_1, Φ_2 are the canonical projections of V_1, V_2 onto Δ, respectively. In the above definition of the family $\mathscr{F}(\mathscr{J}, G)$, analytically equivalent fibre spaces may be considered as the *same* fibre space. More precise classifications of analytic fibre spaces of elliptic curves will be introduced later.

Let α_ρ be a small circle with positive orientation around a_ρ. The restriction $G \mid E_\rho$ of G to a circular neighborhood E_ρ of a_ρ is determined by the representation (S_{α_ρ}) of α_ρ.

THEOREM 7. *The type of the fibre* C_{a_ρ} *of any fibre space* $V \in \mathscr{F}(\mathscr{J}, G)$ *is determined uniquely by* (S_{α_ρ}) (see Table I).

By a holomorphic section of the fibre space V over an open subset

TABLE I

Normal form of (S_{α_ρ})	Type of C_{a_ρ}	$C^\sharp_{a_\rho}$	$\Theta^\sharp_{\rho 0}$	Behavior of $\mathscr{J}(u)$ at a_ρ
1	regular	torus	torus	
$\begin{pmatrix} -1 & 0 \\ 0 & -1 \end{pmatrix}$	I_0^*	$C \times Z_2 \times Z_2$	C	regular
$\begin{pmatrix} 1 & b \\ 0 & 1 \end{pmatrix}$	I_b	$C^* \times Z_b$	C^*	
$\begin{pmatrix} -1 & -b \\ 0 & -1 \end{pmatrix}$	I_b^*	$C \times Z_2 \times Z_2$	C	pole of order b
$\begin{pmatrix} 1 & 1 \\ -1 & 0 \end{pmatrix}$	II	C	C	
$\begin{pmatrix} 0 & -1 \\ 1 & 1 \end{pmatrix}$	II*	C	C	$\mathscr{J}(a_\rho) = 0$
$\begin{pmatrix} 0 & 1 \\ -1 & 0 \end{pmatrix}$	III	$C \times Z_2$	C	
$\begin{pmatrix} 0 & -1 \\ 1 & 0 \end{pmatrix}$	III*	$C \times Z_2$	C	$\mathscr{J}(a_\rho) = 1$
$\begin{pmatrix} 0 & 1 \\ -1 & -1 \end{pmatrix}$	IV	$C \times Z_3$	C	
$\begin{pmatrix} -1 & -1 \\ 1 & 0 \end{pmatrix}$	IV*	$C \times Z_3$	C	$\mathscr{J}(a_\rho) = 0$

Z_b denotes the cyclic group of order b.

$E \subset \Delta$ we mean a holomorphic map $\varphi : u \to \varphi(u)$ of E into V such that $\Phi(\varphi(u)) = u$.

THEOREM 8. *The family $\mathscr{F}(\mathscr{J}, G)$ contains one and only one fibre space B which has a global holomorphic section over Δ.*

We note that the fibre space B is an *algebraic surface.*

Let $\mathscr{G}$ and M be complex manifolds. We say that $\mathscr{G}$ is an analytic fibre space over M free from singular fibres if there is a holomorphic map Ψ of $\mathscr{G}$ onto M such that, at each point on $\mathscr{G}$, the rank of the Jacobian of Ψ is equal to the dimension of M.

DEFINITION 2. *Let $\mathscr{G}$ be an analytic fibre space over M free from singular fibres such that each fibre $\mathscr{G}_u = \Psi^{-1}(u)$, $u \in M$, is a complex Lie group whose complex structure is that of the complex submanifold $\mathscr{G}_u$ of $\mathscr{G}$. Let $\mathscr{D}$ be the complex submanifold of $\mathscr{G} \times \mathscr{G}$ consisting of all points $(g_1, g_2) \in \mathscr{G} \times \mathscr{G}$ satisfying $\Psi(g_1) = \Psi(g_2)$. $\mathscr{G}$ will be called an analytic fibre system of complex Lie groups over M if the following two conditions are satisfied:* (i) *$u \to 1_u$ is a holomorphic map of M into $\mathscr{G}$, where 1_u denotes the unit of $\mathscr{G}_u$,* (ii) *$(g_1, g_2) \to g_1 \cdot g_2^{-1}$ is a holomorphic map of $\mathscr{D}$ onto $\mathscr{G}$.*

Now we consider the fibre space $B \in \mathscr{F}(\mathscr{J}, G)$. We denote the canonical projection $B \to \Delta$ by Ψ and one of the global holomorphic sections of B by $o : u \to o(u)$. Each fibre $C_u = \Psi^{-1}(u)$, $u \in \Delta'$, of B is a complex torus. The canonical additive group structure on a complex torus is determined uniquely by giving the position of the unit 0. We choose the additive group structure on each complex torus C_u such that the unit 0 coincides with $o(u)$. Then the subspace $B' = B \mid \Delta' = \Psi^{-1}(\Delta')$ of B forms an *analytic fibre system of complex Lie groups over* Δ' in the sense of the above Definition 2.

We denote by $B^{\#}$ the open subset of B consisting of all points z satisfying

$$\left|\frac{\partial \tau_a(\Psi(z))}{\partial z_1}\right|^2 + \left|\frac{\partial \tau_a(\Psi(z))}{\partial z_2}\right|^2 > 0$$

(compare (1)). Clearly we have

$$B^{\#} = \bigcup_\rho C^{\#}_{a_\rho} \cup B', \qquad B' = B \mid \Delta',$$

where

$$C^{\#}_{a_\rho} = B^{\#} \cap \Psi^{-1}(a_\rho).$$

The fibre $C^{\#}_{a_\rho}$ is written in the form

$$C^{\#}_{a_\rho} = \bigcup_{n_{\rho s}=1} \Theta^{\#}_{\rho s}, \qquad \Theta^{\#}_{\rho s} = B^{\#} \cap \Theta_{\rho s},$$

where $\bigcup_{n_{\rho s}=1}$ denotes the sum extended over all s for which $n_{\rho s} = 1$. If C_{a_ρ} is regular, then $C^{\#}_{a_\rho} = C_{a_\rho}$; otherwise each component $\Theta^{\#}_{\rho s}$ of $C^{\#}_{a_\rho}$ is (analytically homeomorphic to) either $\mathbf{C}$ or $\mathbf{C}^*$.

PROPOSITION 1. *The analytic fibre space $B^\#$ over Δ has a unique structure of analytic fibre system of complex abelian Lie groups which is an extension of the structure of analytic fibre system of complex Lie groups on B'. The group structures of the fibres $C^\#_{a_\rho}$ of $B^\#$ are given in* Table I.

Let E be an arbitrary open domain on Δ and let $\varphi_\nu : u \to \varphi_\nu(u)$, $\nu = 1, 2$, be holomorphic sections of $B^\#$ over E. Then, by Proposition 1, $u \to \varphi_1(u) \pm \varphi_2(u)$ is also a holomorphic section of $B^\#$ over E, where $+$ or $-$ denotes the group operation on each fibre of $B^\#$. Thus the space $\Gamma(B^\# \mid E)$ of holomorphic sections of $B^\#$ over E forms an additive group in a canonical manner and therefore *the sheaf over Δ of germs of holomorphic sections of $B^\#$ is defined. We denote this sheaf by $\Omega(B^\#)$.*

For any holomorphic section $\varphi : u \to \varphi(u)$ of $B^\#$ over E,

$$L^\#(\varphi) : z \to z + \varphi(\Psi(z))$$

is a fibre preserving analytic homeomorphism of $B^\# \mid E$ onto $B^\# \mid E$.

PROPOSITION 2. *$L^\#(\varphi) : B^\# \mid E \to B^\# \mid E$ can be extended uniquely to a fibre preserving analytic homeomorphism $L(\varphi)$ of $B \mid E$ onto $B \mid E$.*

In view of this proposition, $\Omega(B^\#)$ may be regarded as a *sheaf of structure groups* acting on the fibre space B (compare Grothendieck [4]).

We consider the first cohomology $H^1(\Delta, \Omega(B^\#))$ of Δ with coefficients in the sheaf $\Omega(B^\#)$. Given an element $\sigma \in H^1(\Delta, \Omega(B^\#))$, let $\{E_j\}$ be a sufficiently fine covering of Δ and let $\{\sigma_{jk}\}$, $\sigma_{jk} \in \Gamma(B^\# \mid E_j \cap E_k)$, be a 1-cocycle on the nerve of the covering $\{E_j\}$ representing the cohomology class σ.

DEFINITION 3. *B^σ is the fibre space of elliptic curves over Δ obtained from the collection $\{B \mid E_j\}$ of the pieces $B \mid E_j$ of the fibre space B by identifying $z_j \in B \mid E_j$ and $z_k \in B \mid E_k$ if and only if $z_j = L(\sigma_{jk}) \cdot z_k$.*

It is clear that B^σ is determined uniquely (up to an analytic equivalence) by the cohomology class σ of the 1-cocycle $\{\sigma_{jk}\}$.

THEOREM 9. *The family $\mathscr{F}(\mathscr{J}, G)$ consists of all fibre spaces B^σ, $\sigma \in H^1(\Delta, \Omega(B^\#))$.*

We denote by $\Theta^\#_{\rho 0}$ the component of $C^\#_{a_\rho}$ containing the unit $o(a_\rho)$ of $C^\#_{a_\rho}$ and let

$$B^\#_0 = \bigcup_\rho \Theta^\#_{\rho 0} \cup B'.$$

$B^\#_0$ has also a structure of an analytic fibre system of complex Lie groups which is the restriction of the structure of $B^\#$ and therefore $\Omega(B^\#_0)$ is a subsheaf of $\Omega(B^\#)$. The factor sheaf $F = \Omega(B^\#)/\Omega(B^\#_0)$ has the following property:

$$F_u = \begin{cases} 0, & \text{for } u \neq a_\rho, \\ \text{finite abelian group}, & \text{for } u = a_\rho, \end{cases}$$

where F_u denotes the stalk of F over u. Hence we obtain the exact sequence

(3) $$\dots \to H^0(\Delta, F) \to H^1(\Delta, \Omega(B_0^\sharp)) \to H^1(\Delta, \Omega(B^\sharp)) \to 0.$$

For an element σ of $H^1(\Delta, \Omega(B_0^\sharp))$ we define B^σ in the same manner as in Definition 3. In view of (3), we obtain from Theorem 9 the following

THEOREM 10. *The family $\mathscr{F}(\mathscr{J}, G)$ consists of all fibre spaces B^σ, $\sigma \in H^1(\Delta, \Omega(B_0^\sharp))$.*

The notions of fibre spaces and their equivalences depend on the sheaf of structure groups (compare Grothendieck [4]). We mean by a Ξ-fibre space a fibre space with the sheaf of structure groups Ξ and we say that two fibre spaces are Ξ-equivalent if they are equivalent as Ξ-fibre spaces. $V = B^\sigma$, $\sigma \in H^1(\Delta, \Omega(B_0^\sharp))$, may be considered as an analytic fibre space, as an $\Omega(B^\sharp)$-fibre space, or as an $\Omega(B_0^\sharp)$-fibre space, and the element $\sigma \in H^1(\Delta, \Omega(B_0^\sharp))$ represents the $\Omega(B_0^\sharp)$-equivalence class of B^σ.

4. Clearly the fibre $C_{0u}^\sharp$ of $B_0^\sharp$ over u is given by

$$C_{0u}^\sharp = \begin{cases} C_u, & \text{for } u \in \Delta', \\ \Theta_{\rho 0}^\sharp, & \text{for } u = a_\rho, \end{cases}$$

where $\Theta_{\rho 0}^\sharp$ is isomorphic to $\mathbf{C}$, $\mathbf{C}^*$, or a complex torus if C_{a_ρ} is regular (compare Table I). Let $\mathfrak{f}_u$ be the tangent space of $C_{0u}^\sharp$ at $o(u)$. Then

$$\mathfrak{f} = \bigcup_{u \in \Delta} \mathfrak{f}_u$$

forms a *complex line bundle* over Δ in a canonical manner. Since $\mathfrak{f}_u$ may be regarded as the infinitesimal group of the complex Lie group $C_{0u}^\sharp$ which is isomorphic to a complex torus, $\mathbf{C}$, or $\mathbf{C}^*$, we have a canonical homomorphism

$$h_u : \mathfrak{f}_u \to C_{0u}^\sharp.$$

It follows that there is a locally biregular holomorphic map

$$h : \mathfrak{f} \to B_0^\sharp$$

whose restriction to each fibre $\mathfrak{f}_u$ coincides with h_u. Clearly h induces a homomorphism: $\Omega(\mathfrak{f}) \to \Omega(B_0^\sharp)$ of the sheaf $\Omega(\mathfrak{f})$ of holomorphic sections of $\mathfrak{f}$ onto the sheaf $\Omega(B_0^\sharp)$. We denote this homomorphism by the same symbol h.

PROPOSITION 3. *We have the exact sequence*

(4) $$0 \to G \to \Omega(\mathfrak{f}) \xrightarrow{h} \Omega(B_0^\sharp) \to 0,$$

where G is the homological invariant of the fibre space B.

We obtain from (4) the corresponding *exact cohomology sequence*

(5) $$\dots \to H^1(\Delta, G) \to H^1(\Delta, \Omega(\mathfrak{f})) \xrightarrow{h^*} H^1(\Delta, \Omega(B_0^\sharp)) \xrightarrow{\delta^*} H^2(\Delta, G) \to 0.$$

We define

$$c(\sigma) = \delta^*(\sigma), \qquad \text{for } \sigma \in H^1(\Delta, \Omega(B_0^\sharp)).$$

Let $A(\mathfrak{f})$ and $A(B_0^\sharp)$ be the sheaves over Δ of differentiable sections of $\mathfrak{f}$ and $B_0^\sharp$, respectively. $A(B_0^\sharp)$ is a sheaf of structure groups acting on B considered as a *differentiable* fibre space. From the exact commutative diagram

$$\begin{array}{ccccccccc} 0 & \rightarrow & G & \rightarrow & \Omega(\mathfrak{f}) & \rightarrow & \Omega(B_0^\sharp) & \rightarrow & 0 \\ & & \| & & \downarrow\iota & & \downarrow\iota & & \\ 0 & \rightarrow & G & \rightarrow & A(\mathfrak{f}) & \rightarrow & A(B_0^\sharp) & \rightarrow & 0, \end{array}$$

where ι denotes the inclusion map, we obtain the exact commutative diagram

$$\begin{array}{ccccccc} \rightarrow H^1(\Delta, \Omega(\mathfrak{f})) & \rightarrow & H^1(\Delta, \Omega(B_0^\sharp)) & \rightarrow & H^2(\Delta, G) & \rightarrow & 0 \\ \downarrow & & \downarrow\iota^* & & \| & & \\ 0 & \longrightarrow & H^1(\Delta, A(B_0^\sharp)) & \rightarrow & H^2(\Delta, G) & \rightarrow & 0 \end{array}$$

This shows that $c(\sigma)$ represents the $A(B_0^\sharp)$-equivalence class of the fibre space B^σ over Δ. In particular we have

THEOREM 11. *If $c(\sigma) = c(\eta)$, then the fibre spaces B^σ and B^η are differentiably equivalent.*

We call $c(\sigma)$ the *characteristic class* of the $\Omega(B_0^\sharp)$-fibre space B^σ over Δ.

By a *complex analytic family* of compact complex manifolds we mean the family $\{V_t \mid t \in M\}$ of fibres $V_t = \pi^{-1}(t)$ of an analytic fibre space $\mathscr{V}$ free from singular fibres over a *connected* complex manifold M whose fibres are compact, where π is the canonical projection $\mathscr{V} \rightarrow M$ (see Kodaira and Spencer [8]). Moreover we say that a compact complex manifold W is a (complex analytic) *deformation* of V if W and V belong to one and the same complex analytic family. It is clear that any deformation W of V is differentiably homeomorphic to V.

THEOREM 12. *$\{B^{\sigma+h^*(t)} \mid t \in H^1(\Delta, \Omega(\mathfrak{f}))\}$ forms a complex analytic family of compact complex manifolds, where σ is an element of $H^1(\Delta, \Omega(B_0^\sharp))$.*

We remark here that $H^1(\Delta, \Omega(\mathfrak{f}))$ has a canonical structure of complex vector space. We set

$$\mathscr{V}^{(c)} = \{B^{\sigma+h^*(t)} \mid t \in H^1(\Delta, \Omega(\mathfrak{f}))\}, \qquad c = c(\sigma).$$

In view of (5), the family $\mathscr{V}^{(c)}$ is composed of all fibre spaces B^η, $c(\eta) = c$, with repetitions. Hence we have

THEOREM 13. *If $c(\eta) = c(\sigma)$, then B^η is a deformation of B^σ.*

It is clear that the projection $\Phi : B^\sigma \rightarrow \Delta$ induces an isomorphism Φ^* of $\mathscr{M}(\Delta)$ into $\mathscr{M}(B^\sigma)$ and therefore $\dim \mathscr{M}(B^\sigma) \geq 1$. Using the fact that, if $\dim \mathscr{M}(B^\sigma) = 1$, B^σ contains no irreducible curve other than the components of the fibres (see Theorem 2), we obtain the following

THEOREM 14. *B^σ is an algebraic surface if and only if σ is an element of finite order of $H^1(\Delta, \Omega(B_0^\sharp))$.*

Combining this with Theorem 13, we get

THEOREM 15. *If $c(\sigma)$ is an element of finite order of $H^2(\Delta, G)$, then B^σ is a deformation of an algebraic surface.*

5. We denote by $c(\mathfrak{f})$ the characteristic class of the complex line bundle $\mathfrak{f}$ over Δ. In view of the canonical isomorphism: $H^2(\Delta, \mathbf{Z}) \cong \mathbf{Z}$, we may suppose that $c(\mathfrak{f})$ is an integer. Moreover we denote by $\nu(T)$ the number of the singular fibres of B of type T and by j the order of the meromorphic function $\mathscr{J}(u)$, i.e. the total multiplicities of the zeros of $\mathscr{J}(u)$.

PROPOSITION 4. *$\delta = -c(\mathfrak{f})$ is given by*

$$12\delta = j + 6\textstyle\sum_b \nu(\mathrm{I}_b^*) + 2\nu(\mathrm{II}) + 10\nu(\mathrm{II}^*) + 3\nu(\mathrm{III}) + 9\nu(\mathrm{III}^*) + 4\nu(\mathrm{IV}) + 8\nu(\mathrm{IV}^*).$$

This proposition shows, in particular, that $c(\mathfrak{f}) \leqq 0$ and that $c(\mathfrak{f}) = 0$ if and only if B has no singular fibre and $\mathscr{J}(u)$ is reduced to a constant. By using Riemann-Roch's theorem we obtain

$$\dim H^1(\Delta, \Omega(\mathfrak{f})) = \begin{cases} p - 1 + \delta, & \text{if } \mathfrak{f} \neq 0, \\ p, & \text{if } \mathfrak{f} = 0, \end{cases} \tag{6}$$

where p is the genus of Δ.

PROPOSITION 5. *$H^2(\Delta, G)$ is a finite abelian group, provided that G is not trivial.*

It is clear that, if G is the trivial sheaf $\mathbf{Z} \oplus \mathbf{Z}$, $H^2(\Delta, G)$ is isomorphic to $\mathbf{Z} \oplus \mathbf{Z}$.

Now we consider an arbitrary fibre space $V = B^\sigma$, $\sigma \in H^1(\Delta, \Omega(B_0^\sharp))$, belonging to $\mathscr{F}(\mathscr{J}, G)$. Let $\mathfrak{k}$ be the canonical bundle of Δ.

THEOREM 16. *The canonical bundle K of V is induced from $\mathfrak{k} - \mathfrak{f}$ by the projection Φ of V onto $\Delta : K = \Phi^*(\mathfrak{k} - \mathfrak{f})$.*

Combining this with (6), we infer that the geometric genus $p_g(V)$ is given by

$$p_g(V) = \begin{cases} p - 1 + \delta, & \text{if } \mathfrak{f} \neq 0, \\ p, & \text{if } \mathfrak{f} = 0. \end{cases}$$

THEOREM 17. *The Euler number $c_2(V)$ of V is given by $c_2(V) = 12\delta$.*

An analytic fibre space V of (elliptic) curves over Δ is called an analytic *fibre bundle* if each point $u \in \Delta$ has a neighborhood E_u such that $V \mid E_u = E_u \times C_u$ in the *complex analytic sense*. It can be shown that, if G is trivial, B is analytically trivial i.e. $B = \Delta \times C_0$ in the complex analytic sense and therefore $V = B^\sigma$ is an analytic principal bundle of elliptic curves, i.e. an analytic fibre bundle over Δ whose fibre is a non-singular elliptic curve C_0 and whose structure group is the translation group acting on C_0. We infer from Proposition 5 and Theorem 14 that, in case G is not trivial, *each complex analytic family $\mathscr{V}^{(c)}$ contains an algebraic surface.* Hence we obtain

THEOREM 18. *An analytic fibre space V of elliptic curves over a curve Δ*

free from exceptional curves and from multiple singular fibres is a deformation of an algebraic surface unless V is an analytic principal bundle of elliptic curves.

Since the group $H^1(\Delta, G)$ is finitely generated, we infer from (5) and Theorem 14 the following

THEOREM 19. *If G is not trivial and if* $\dim H^1(\Delta, \Omega(\mathfrak{f})) \geqq 1$, *the general member of each complex analytic family $\mathscr{V}^{(c)}$ is not algebraic, or, more precisely, the member $B^{\sigma+h^*(t)}$ of $\mathscr{V}^{(c)}$ is algebraic if and only if t belongs to a countable subset of $H^1(\Delta, \Omega(\mathfrak{f}))$.*

In case G is trivial, each family $\mathscr{V}^{(c)}$, $c \neq 0$, contains no algebraic surface. The family $\mathscr{V}^{(0)}$ contains the algebraic surface $B = \Delta \times C_0$, but, if $\dim H^1(\Delta, \Omega) \geqq 1$, the general member of $\mathscr{V}^{(0)}$ is not algebraic.

6. Now we consider an arbitrary compact analytic surface V with $\dim \mathscr{M}(V) = 1$. By Theorems 3 and 2, V is a quadratic transform of an analytic fibre space W of elliptic curves over a curve Δ_0 containing no exceptional curve. By Theorem 6, W is represented in the form $W = \tilde{W}/\mathfrak{G}$, where $\tilde{W}$ is an analytic fibre space of elliptic curves over a finite ramified covering $\tilde{\Delta}_0$ of Δ_0 free from multiple singular fibres. Let $\Delta = \tilde{\Delta}_0$. Then, by Theorem 10, W has the form B^σ, $\sigma \in H^1(\Delta, \Omega(B_0^\sharp))$. Thus we obtain

THEOREM 20. *Every compact analytic surface V with* $\dim \mathscr{M}(V) = 1$ *is represented in the form $V = \ldots Q_{p_3} Q_{p_2} Q_{p_1}(B^\sigma/\mathfrak{G})$, $\sigma \in H^1(\Delta, \Omega(B_0^\sharp))$, where $\mathfrak{G}$ is a finite abelian group of analytic automorphisms of B^σ introduced in Theorem* 6.

An analytic fibre space V of elliptic curves over Δ containing no exceptional curve will be called an *analytic quasi-bundle* of elliptic curves if each singular fibre C_{a_ρ} of V is of type ${}_m\mathrm{I}_0$, m depending on ρ, and if $V \mid \Delta'$ is an analytic principal bundle of elliptic curves over $\Delta' = \Delta - \{a_\rho\}$. It can be shown that $W = \tilde{W}/\mathfrak{G}$ is an analytic quasi-bundle of elliptic curves over Δ_0 if and only if $\tilde{W} = B^\sigma$ is an analytic principal bundle of elliptic curves over $\tilde{\Delta}_0 = \Delta$.

Suppose that $\tilde{W} = B^\sigma$ is not an analytic principal bundle of elliptic curves. Let X be the underlying differentiable manifold on which the complex structure B^σ is defined and consider $\mathfrak{G}$ as a group of differentiable automorphisms of X. By a detailed analysis of the structures of the automorphisms of $\mathfrak{G}$, we infer that there exists a linear subspace M of $H^1(\Delta, \Omega(\mathfrak{f}))$ satisfying the following two conditions: (i) The complex analytic family $\{B^{\sigma+h^*(t)} \mid t \in M\}$ contains an algebraic surface, (ii) The automorphisms of $\mathfrak{G}$ are biregular with respect to each complex analytic structure $B^{\sigma+h^*(t)}$, $t \in M$, defined on X. Letting

$$V_t = \ldots Q_{p_3} Q_{p_2} Q_{p_1}(B^{\sigma+h^*(t)}/\mathfrak{G}),$$

we obtain a complex analytic family $\{V_t \mid t \in M\}$ of deformations V_t of $V_0 = V$ which contains an algebraic surface. Thus we obtain

THEOREM 21. *A compact analytic surface V with* dim $\mathcal{M}(V) \geqq 1$ *is a deformation of an algebraic surface unless V is a quadratic transform of an analytic quasi-bundle of elliptic curves over a curve.*

In case $\tilde{W} = B^{\sigma}$ is an analytic principal bundle of elliptic curves, it can be shown that, if $V = \dots Q_{p_2} Q_{p_1}(B^{\sigma}/\mathfrak{G})$ has a Kähler metric, V is a deformation of an algebraic surface. Combining this with Theorem 21, we obtain

THEOREM 22. *Every compact Kähler surface having at least one non-constant meromorphic function is a deformation of an algebraic surface.*

Now we consider a compact Kähler surface V with dim $\mathcal{M}(V) = 0$. If the first Betti number $b_1(V)$ of V is positive, then, by Theorem 4, V is a quadratic transform of a complex torus, and therefore V is a deformation of an algebraic surface. Combining this with the above Theorem 22, we obtain

THEOREM 23. *Every compact Kähler surface V with the first Betti number $b_1(V) > 0$ is a deformation of an algebraic surface.*

REFERENCES

[1] G. CASTELNUOVO and F. ENRIQUES, Sopra alcune questioni fondamentali nella teoria delle superficie algebriche, *Annali di Math., Ser. III, 6* (1901), pp. 165–225.

[2] W. L. CHOW, On complex analytic varieties, to appear in the *Amer. J. Math.*

[3] W. L. CHOW and K. KODAIRA, On analytic surfaces with two independent meromorphic functions, *Proc. Nat. Acad. Sciences 38* (1952), pp. 319–325.

[4] A. GROTHENDIECK, *A general theory of fibre spaces with structure sheaf*, University of Kansas, 1955.

[5] H. HOPF, Schlichte Abbildungen und lokale Modificationen 4-dimensionaler komplexer Mannigfaltigkeiten, *Comm. Math. Helv. 29* (1955), pp. 132–156.

[6] K. KODAIRA, *On compact complex analytic surfaces I*, to appear in *Ann. of Math.*

[7] ———, On Kähler varieties of restricted type, *Ann. of Math. 60* (1954), pp. 28–48.

[8] K. KODAIRA and D. C. SPENCER, On deformations of complex analytic structures, I–II, *Ann. of Math.* 67 (1958) pp. 328–466.

[9] C. L. SIEGEL, Meromorphe Funktionen auf kompakten analytischen Mannigfaltigkeiten, *Nachr. Akad. Wiss.* Göttingen (1955), pp. 71–77.

[10] O. ZARISKI, Algebraic surfaces, *Ergeb. Math. Berlin* (1935).

The Conformal Mapping of Riemann Surfaces

Maurice Heins
UNIVERSITY OF ILLINOIS

1. Introduction

THE object of this essay is to examine the status of the chapter of the theory of Riemann surfaces which is concerned with conformal maps of Riemann surfaces. It is to be understood that we take the term in its general sense and do not require the considered maps to be *univalent onto* maps. We set as our goal a study of the following problems: (1) the relation between conformal mapping theory so conceived and other chapters of the general theory of Riemann surfaces, (2) the development of an autonomous mapping theory in which conformal maps are studied for their own sake, (3) the application of such an autonomous theory to problems of classical analysis. The rapid development of Riemann surface theory at the present time underscores eloquently the fact that we use the word "essay" fully conscious of its etymological force.

When we ask what other areas of Riemann surface theory show a close relation with mapping theory, two come to mind. One is the venerable theory of *analytische Gebilde* of Weierstrass which affords an important realization of abstract Riemann surfaces. Here, as we shall see, the language and methods of mapping theory afford a natural approach to the classical realization problems which go back to Riemann and Koebe as well as to allied realization problems where more stringent conditions are imposed. It will be apparent that in these considerations the mapping theory serves as a bridge between certain entities associated with analytische Gebilde, and their generalizing counterparts in the abstract theory.

A second area which is intimately connected with mapping theory is the study of the algebraic structures associated with a Riemann surface—the field of meromorphic functions on the given surface and its algebraic substructures (subfields and subrings). The relation between conformal equivalence of Riemann surfaces and the isomorphism (with preservation of constants) of the associated fields of meromorphic functions is classical in the compact case. In the non-compact case the pioneer investigations of

Chevalley and Kakutani [10] in the early Forties have been an impetus to a number of studies which are concerned with the interrelation of algebraic structure and conformal mapping. We shall comment on this work later.

1. AN AUTONOMOUS MAPPING THEORY. When we seek to develop mapping theory for its own sake, we are confronted with the fact that we already have a more or less developed mapping theory in the plane case—the study of meromorphic functions with a given plane domain which take values in a second plane domain. In certain important cases the theory has taken a highly precise form as, for example, in the Nevanlinna theory of meromorphic functions. When we turn to the problem of cultivating a mapping theory for Riemann surfaces which is freed of the restriction of planarity, we face the question where to place the emphasis. Various avenues are open. For one we have the topological approach in which the notion of an interior transformation due to Stoilow dominates. We are then concerned with the study of the purely topological characteristics of interior transformations. Reference should be made to the work of G. T. Whyburn [24], M. Morse [14], Jenkins, Morse and Jenkins, Morse and Heins. Antecedent work on the theory of coverings and the investigations of L. E. J. Brouwer on finite groups of homeomorphisms of compact surfaces onto themselves belong most appropriately to this domain. There is next an intermediate domain where metric notions dominate—the Ahlfors' theory of covering surfaces [1], which is the root source of a large number of investigations, and the theory of quasi-conformal maps actively cultivated at the present time.

The theory of conformal maps of Riemann surfaces may be conceived as a specialized substratum of these more general theories. On the other hand, we may approach the theory on its own terms and seek to develop it in terms of methods which are characteristic for it, in particular potential-theoretic methods. It is this aspect of the question which we propose to examine here in greater detail. It is inevitable that there should be lurking in the background the problem of whether the results achieved in such a framework are special to the conformal theory or not. Further in the construction of such a theory it is natural that we should seek to embrace as far as possible the classical plane theory as well as to differentiate between the phenomena peculiar to the case of surfaces of infinite genus and those of surfaces of finite genus.

A method which has been of service in the study of conformal maps of Riemann surfaces is that which we term "*the method of the Lindelöf Principle*" [9a]. In brief, it consists in the examination of the action of a conformal mapping on the Green's function and correlating this action with the covering properties of the mapping. As its name suggests, its source is to be traced to the Lindelöf Principle of the classical theory.

Other methods are important as well—for one, that of uniformization. We restrict our attention to the Lindelöf method in order to show that it affords a language and access for the study of a number of diverse problems. We shall see that this method leads naturally to a local classification of a conformal map at each point of the range surface whose embodiment is contained in the notion of a map's being of "type-$B\ell$ at a point" (§4).

Reference should be made to the investigations of Lehto on functions of bounded characteristic [12] and of Parreau on the general theory of conformal maps of Riemann surfaces [17b] which are contemporary with those of the author. In these investigations the Lindelöfian point of view also comes to the fore. Among other recent investigations directed toward mapping theory mention should be made of the work of Fourès [6a, b] and the Japanese school, in particular, that of Ohtsuka [16].

2. Applications to the classical theory. Here we have in mind, in part, a reexamination of certain problems of classical function-theory from the point of view of the mapping theory just referred to. Among other results which such a renewed study brings, we mention a sharpened form of the meromorphic case of the Denjoy-Carleman-Ahlfors Theorem, whose formulation is directly based on Lindelöfian notions [9c]. There exist non-trivial examples of meromorphic functions whose growth is measured exactly by the sharpened Denjoy-Carleman-Ahlfors Theorem but not by the original form of the theorem.

There is an area of the classical nineteenth century function-theory which appears to be practically unexplored. We refer to the mapping properties of certain distinguished "multiple-valued" meromorphic functions—namely, abelian integrals and solutions of linear homogeneous differential equations. It will be seen that the Lindelöfian approach will permit us to come to grips with such questions in a number of important cases. The work reported in the last sections of this essay cannot be regarded as final, but it does show that a study of the mapping properties of the classical "multiple-valued" transcendental functions is certainly accessible.

We now turn to a more detailed examination of the above questions.

2. Analytische Gebilde

We recall the notion of an analytic structure in the sense of Weierstrass (= analytische Gebilde). The formulation given here is essentially that of Weyl [23]. We consider the class K of ordered pairs (f, g) where f and g are non-constant meromorphic functions having a common domain consisting of a plane region containing the origin and the map $z \to (f(z), g(z))$ is univalent. We say that members (f_1, g_1) and (f_2, g_2) are *equivalent* provided that there exists a function φ analytic at $z = 0$ and satisfying $\varphi(0) = 0$, such that

$$f_1(z) = f_2(\varphi(z)), \quad g_1(z) = g_2(\varphi(z))$$

for sufficiently small z. This requirement defines an equivalence relation in K. We term an equivalence class of this relation an *element*. The element containing (f, g) will be denoted by $e(f, g)$. We proceed by introducing a topology on the set $\mathfrak{E}$ of elements which renders each component of $\mathfrak{E}$ a 2-dimensional manifold, and thereupon endowing each component of $\mathfrak{E}$ with a conformal structure. These components are the underlying sets of the *analytische Gebilde*.

Given $(f, g) \in K$, we define $V(f, g)$ as the set of elements $e(f^t, g^t)$ where $f^t(z) = f(t + z)$, $g^t(z) = g(t + z)$, $t \in$ domain of f and g. A topology is introduced on $\mathfrak{E}$ by defining O ($\subset \mathfrak{E}$) to be *open* provided that for each $e \in O$ there exists a pair $(f, g) \in e$ such that $V(f, g) \subset O$. This requirement endows $\mathfrak{E}$ with a Hausdorff topology. The map $t \to e(f^t, g^t)$ defines a homeomorphism of the domain of f and g onto $V(f, g)$. Thus $\mathfrak{E}$ is 2-dimensional. Let $\mathfrak{A}$ denote a component of $\mathfrak{E}$. Let $\mathfrak{T}$ denote the set of homeomorphisms $t \to e(f^t, g^t)$ where all (f, g) belonging to the elements $e \in \mathfrak{A}$ are taken into account. Then $\mathfrak{T}$ endows $\mathfrak{A}$ with a conformal structure. The analytische Gebilde are the $\mathfrak{A}$ with conformal structure so defined.

We note that $(f(0), g(0))$ is independent of $(f, g) \in e$. Following the terminology of Saks and Zygmund [22] we term the common value of $f(0)$ the *center* of e and denote it by $c(e)$ and the common value of $g(0)$ the *value* of e and denote it by $v(e)$. With a given analytische Gebilde $\mathfrak{A}$ we associate the two functions

$$c : e \to c(e), \quad v : e \to v(e), \quad e \in \mathfrak{A},$$

termed respectively the *center function* and the *value function* of $\mathfrak{A}$. They are, of course, meromorphic in $\mathfrak{A}$.

The following theorem links the notion of an analytische Gebilde with conformal mapping of Riemann surfaces.

THEOREM 1. *Let F denote a Riemann surface and let f and g denote non-constant meromorphic functions on F. Then there exist an analytische Gebilde $\mathfrak{A}$ and a conformal map φ of F into $\mathfrak{A}$ such that $f = c \circ \varphi$, $g = v \circ \varphi$. There is precisely one pair $(\mathfrak{A}, \varphi)$ for which these conditions are fulfilled.*

Theorem 1 permits us to characterize triples $(\mathfrak{A}, c, v)$ up to a conformal equivalence. Triples (F_k, f_k, g_k), $k = 1, 2$, where the F_k are Riemann surfaces and f_k and g_k are non-constant meromorphic functions on F_k, are said to be *equivalent* provided that there exists a univalent conformal map φ of F_1 onto F_2 such that $f_1 = f_2 \circ \varphi$, $g_1 = g_2 \circ \varphi$. An equivalence is thereby defined. A triple (F_1, f_1, g_1) is said to be *extendible* provided that there exist a triple (F_2, f_2, g_2) and a conformal map φ of F_1 into F_2 which is *not* a univalent map of F_1 *onto* F_2 such that $f_1 = f_2 \circ \varphi$ and $g_1 = g_2 \circ \varphi$. (For an allied notion of extendibility of conformal maps of Riemann surfaces, cf. [4].) Thanks to Theorem 1 it is easy to establish

THEOREM 2. *The triple* (F, f, g) *is equivalent to a triple* $(\mathfrak{A}, c, v)$ *if and only if* (F, f, g) *is not extendible.*

The classical realization problem of Riemann and Koebe may be formulated as follows: Given a Riemann surface F and a non-constant meromorphic function f on F, does there exist a non-constant meromorphic function g on F such that (F, f, g) is equivalent to a triple $(\mathfrak{A}, c, v)$?

If F is compact and q is a point of F at which the multiplicity of f, $n(q; f)$, is 1, on choosing, as in the traditional argument, a meromorphic function g on F which takes distinct values at the distinct antecedents of $f(q)$ with respect to f, we see that (F, f, g) is not extendible, for, if this were not the case, the extending map φ would have constant valence† and $\varphi(p) = \varphi(q)$ would imply that $p = q$ so that φ would be univalent onto.

From our point of view the non-compact case is more interesting.‡ An examination of the standard proofs of the Koebe realization theorem reveals that they center around the construction of a g which renders φ of Theorem 1 a univalent conformal map of F onto $\mathfrak{A}$. We go further and ask whether it is possible to construct an $\mathfrak{A}$ subject to even more stringent conditions, e.g. with a pole-free *value* function. That this is the case and that we may construct an $\mathfrak{A}$ with the value function subject to certain restrictions as far as its covering properties are concerned may be seen as follows.

Again let $q \in F$ satisfy $n(q; f) = 1$. There exists an analytic function g on F which vanishes at q but at no other antecedent of $f(q)$ with respect to f and which has the following properties: (1) for each bounded plane region Ω each component of $g^{-1}(\Omega)$ is relatively compact, (2) for each compact set $K \subset F$, each component of $F - K$ which is not relatively compact contains a zero of g. The existence of such a g is assured by a construction of the type of [9f] which is based upon the fundamental approximation theorem of Behnke and Stein [2] and Pick-Nevanlinna interpolation methods for bounded analytic functions on Riemann surfaces having finite topological characteristics (genus and contour number). We are now in a position to take advantage of the special covering properties of g.

Suppose that φ is a conformal map of F into a Riemann surface F_1 and f_1 and g_1 are meromorphic functions on F_1 such that $f = f_1 \circ \varphi$ and

† Explicitly, the *valence*, ν_φ, of a conformal map $\varphi : F \to G$, is the function with domain G defined by $\nu_\varphi(r) = \sum_{\varphi(p)=r} n(p; \varphi)$. It is lower semi-continuous on G.

‡ For a modern treatment of the question of the existence of an $\mathfrak{A}$ conformally equivalent to a given non-compact F, cf. the recent monograph of Pfluger [18].

$g = g_1 \circ \varphi$. We first observe that g transmits its covering property to φ: for each region $\Omega \subset \varphi(F)$ where Ω is relatively compact (relative to $\varphi(F)$), each component of $\varphi^{-1}(\Omega)$ is relatively compact. Next we note that if $\varphi(p) = \varphi(q)$, then $f(p) = f(q)$ and $g(p) = g(q)$ so that $p = q$. Since $n(q; f) = 1$, $n(q; \varphi) = 1$. It follows that the valence of φ at $\varphi(q)$ is 1. We conclude that φ is univalent. If $\varphi(F) \neq F_1$, then each frontier point of $\varphi(F)$ would be a pole of g_1. It follows from property (2) of g, that $\varphi(F) = F_1$. Hence (F, f, g) is not extendible.

3. Algebraic structure

As we have remarked in §1, it was the work of Chevalley and Kakutani which first treated the relation between conformal equivalence and isomorphism of allied algebraic structures in the non-compact case. They showed that two plane regions (subject to certain regularity conditions) were conformally equivalent if there existed an isomorphism between the rings of bounded analytic functions on the two regions which preserved the complex constants. Among subsequent investigations the work of L. Bers [3] stands out for its simplicity and elegance. The principal result of Bers is: If φ is an isomorphism of the ring of analytic functions on a region Ω_2 of the finite plane, say $\mathfrak{R}(\Omega_2)$, onto the ring of analytic functions on a region Ω_1 of the finite plane, say $\mathfrak{R}(\Omega_1)$, then either there exists a univalent conformal map ψ_1 of Ω_1 onto Ω_2 such that $\varphi(g) = g \circ \psi_1$, $g \in \mathfrak{R}(\Omega_2)$ or a univalent anticonformal map ψ_2 of Ω_1 onto Ω_2 such that $\varphi(g) = \overline{g \circ \psi_2}$, $g \in \mathfrak{R}(\Omega_2)$. No requirements are imposed on φ beyond that of being a ring isomorphism.

More recently the work of Rudin [21] has shown the relation between conformal equivalence of non-compact Riemann surfaces and isomorphism (with preservation of constants) of the rings of analytic functions on them. Investigations of Royden [20] and the author [9e] have focused attention on the relation between conformal maps of Riemann surfaces and homomorphisms or isomorphisms of associated algebraic structures, where the algebraic applications are *into* and not necessarily *onto*. In these studies, the analytical apparatus which is furnished by the extension of the Weierstrass product theorem to non-compact Riemann surfaces which is due to H. Florack [5] plays an important role.

It is of interest to note that each subfield (containing but not reducing to the constants) of the field of meromorphic functions on a given Riemann surface gives rise to a conformal mapping whose domain is the given surface and which affords a useful representation theorem for the given field.

To be specific, let F denote a given Riemann surface, let $\mathfrak{M}(F)$ denote the field of meromorphic functions on F and let K denote a subfield which contains non-constant members. For each $p \in F$, we introduce the map

$$h_p : f \to f(p), \qquad f \in K.$$

The subfield K is said to *separate* F provided that the map $p \to h_p$, $p \in F$, is univalent.† The notion of a separating field is an important one. Of course, $\mathfrak{M}(F)$ is separating.

Now it turns out that a given subfield K (containing but not reducing to the constants) of $\mathfrak{M}(F)$ is algebraically equivalent to a separating subfield of some $\mathfrak{M}(G)$ and in fact the isomorphism in question may be represented through the mediation of a conformal map of F onto G. We have [9e]

THEOREM 3. *There exist a conformal map ψ of F onto a Riemann surface G and a subfield L of $\mathfrak{M}(G)$ such that L separates G and $g \to g \circ \psi$ maps L onto K. (This map is, of course, an isomorphism which preserves the constants). Further, the triple (G, L, ψ) is determined up to a conformal equivalence in the sense that, if (G_1, L_1, ψ_1) is also a triple fulfilling the imposed conditions, then there exists a univalent conformal map θ of G onto G_1 such that $\psi_1 = \theta \circ \psi$ and $g_1 \to g_1 \circ \theta$, $g_1 \in L_1$, maps L_1 onto L.*

The verification of the essential uniqueness of (G, L, ψ) is a simple matter. The desired θ is simply $\{(\psi(p), \psi_1(p)) \mid p \in F\}$. The *existence* aspect of the theorem may be treated as follows. We start with $G = \{h_p \mid p \in F\}$ and endow G with a conformal structure in such a manner that $\psi : p \to h_p$ is conformal. Further for each $f \in K$, $\hat{f} = \{(h_p, f(p)) \mid p \in F\}$ is a meromorphic function on G satisfying $f = \hat{f} \circ \psi$. The set of $\hat{f}$, $f \in K$, constitutes a separating subfield of $\mathfrak{M}(G)$.

1. AN APPLICATION‡ OF THEOREM 3. Let F denote a non-compact Riemann surface which is such that there exists $f_0 \in \mathfrak{M}(F)$ which has bounded valence and satisfies§ $\{\nu_{f_0}(w) = \max \nu_{f_0}\} \in O_{AB}$. Let $\mathfrak{B}$ denote the subset of $\mathfrak{M}(F)$ consisting of the members of bounded valence together with the constants. *Then there exists a conformal map φ of F into a compact Riemann surface H such that $h \to h \circ \varphi$ maps $\mathfrak{M}(H)$ onto $\mathfrak{B}$. The pair (H, φ) is determined up to a conformal equivalence.*

What Theorem 3 does here is to replace a suitably chosen subfield K of $\mathfrak{M}(F)$ which contains $\mathfrak{B}$ by a separating subfield L. Explicitly K is the subfield of $\mathfrak{M}(F)$ generated by the members of $\mathfrak{M}(F)$ which are bounded in the neighborhood of the ideal boundary. Thanks to the choice of K, the O_{AB} hypothesis on f_0 and the separating property of L we are led to the above assertions concerning $\mathfrak{B}$.

† or equivalently, if for each pair $p, q\ (\neq p) \in F$ there exists $f \in K$ such that $f(p) \neq f(q)$.

‡ A detailed account will appear in the volume of Revues des Mathématiques Pures et Appliquées dedicated to Professor S. Stoilow on the occasion of his seventieth birthday.

§ We recall that an open set of the extended plane belongs to O_{AB} provided that there does not exist a non-constant bounded analytic function on it.

4. The Lindelöf principle

The next matter to concern us will be the Lindelöf principle and some of its immediate consequences for mapping theory. For details concerning the developments of the present section cf. [9a, b].

We shall use the notation $\mathfrak{G}_\Omega$ for the Green's function of a hyperbolic Riemann surface Ω (or a hyperbolic subregion Ω of a Riemann surface). Let φ denote a conformal map of a hyperbolic Riemann surface F into a hyperbolic Riemann surface G. For a given $q \in G$, we consider the superharmonic function $p \to \mathfrak{G}_G(\varphi(p), q)$. Now the exact form of the Lindelöf principle asserts that

$$(4.1) \qquad \mathfrak{G}_G(\varphi(p), q) = \Sigma_{\varphi(r)=q}\, n(r; \varphi)\mathfrak{G}_F(p, r) + u_q(p)$$

where u_q is a non-negative harmonic function on F. We term u_q the *residual function* corresponding to the pole q. Actually (4.1) is nothing more than a simple manifestation of the Riesz decomposition for positive superharmonic functions. From the point of view of mapping theory interest attaches to the dependence of u_q on the parameter q and to a more refined analysis of u_q as a harmonic function on F. We note that u_q is the greatest harmonic minorant of $p \to \mathfrak{G}_G(\varphi(p), q)$. From this fact, we conclude that for each $p \in F$, $q \to u_q(p)$ is upper semi-continuous on G and thereupon, thanks to the Harnack inequality for non-negative harmonic functions (on Riemann surfaces) that $(p, q) \to u_q(p)$ is upper semi-continuous on $F \times G$.

It will be convenient to introduce the following notions concerning non-negative harmonic functions which are due to Parreau [17a]. A non-negative harmonic function u on a Riemann surface F is termed *quasi-bounded* provided that $u = \lim_{\alpha \to +\infty}$ G.H.M. min $\{u, \alpha\}$. Here "G.H.M." denotes "greatest harmonic minorant". On the other hand u is termed *singular* provided that the only bounded non-negative harmonic function on F dominated by u is 0. As Parreau has shown, every non-negative harmonic function u on F admits a unique representation of the form $v + w$ where v is quasi-bounded and w is singular. Further $v = \lim_{\alpha \to +\infty}$ G.H.M. min $\{u, \alpha\}$.

Let $\mathfrak{G}^\alpha_\Omega = \min \{\mathfrak{G}_\Omega, \alpha\}$, $\alpha > 0$. Let v_q and w_q denote respectively the quasi-bounded and singular components of u_q, let $u^\alpha_q =$ G.H.M. $\mathfrak{G}^\alpha_\Omega(\varphi(p), q)$. Then $(p, q) \to u^\alpha_q(p)$ is continuous on $F \times G$. Since $u^\alpha_q(p)$ is non decreasing in α and $\lim_{\alpha \to +\infty} u^\alpha_q = v_q$, we see that $(p, q) \to v_q(p)$ is lower semi-continuous and consequently that $(p, q) \to w_q(p)$ is upper semi-continuous. Given $q, r \in G$, $\alpha > 0$, there exists $\beta > 0$ such that $\mathfrak{G}^\alpha_G(s, q) \leq \beta\mathfrak{G}_G(s, r)$, $s \in G$. It follows that either $v_q > 0$ for all $q \in G$ or else that $v_q = 0$, $q \in G$. We are led thereby to classify φ according to the alternative which occurs.

We shall say that φ is of *type-Bl* provided that $v_q = 0$, $q \in G$, and we shall say that φ is of *type-$B\ell_1$* provided that $u_q = 0$, $q \in G$. ("$B\ell$" for "Blaschke," since such maps enjoy essential properties of Blaschke products.)

In the special case: $F = G = \{|z| < 1\}$, the maps of type-$B\ell$ are the bounded analytic functions whose Fatou boundary values are p.p. of modulus 1 (functions of type-U). For such functions Frostman has shown that $\{q \mid w_q = 0\}$ is an F_σ of zero capacity [7]. Actually, this result holds for an unrestricted φ.

We may now proceed a step further and *localize* the notions of type-$B\ell$ and type-$B\ell_1$. Suppose now that φ is a conformal map of a Riemann surface F into a Riemann surface G. No restrictions are imposed on F or G. Given $q \in G$, we shall say that φ is of *type-$B\ell(B\ell_1)$ at q* provided that there exists a simply-connected Jordan region Ω, $q \in \Omega$, such that (1) $\varphi^{-1}(\Omega) \neq \emptyset$, (2) for each component ω of $\varphi^{-1}(\Omega)$ the restriction of φ to ω is a map of type-$B\ell(B\ell_1)$ of ω into Ω. The set of points of G at which φ is of type-$B\ell(B\ell_1)$ is open. We shall say that φ is *locally of type-$B\ell(B\ell_1)$* provided that φ is of type-$B\ell(B\ell_1)$ at each point of G.

If G is hyperbolic (so that F is also), φ is of type-$B\ell(B\ell_1)$ if and only if φ is locally of type-$B\ell(B\ell_1)$. We shall have occasion to use this result later.

Conformal maps which are locally of type-$B\ell(B\ell_1)$ have a number of important properties. We cite the following: (a) If φ is locally of type-$B\ell$, then either $\sup \nu_\varphi < +\infty$ and $\{\nu_\varphi(q) < \sup \nu_\varphi\}$ is a closed set of zero capacity or $\{\nu_\varphi(q) < +\infty\}$ is an F_σ of zero capacity. (b) If φ is locally of type-$B\ell_1$, then ν_φ is constant (finite or not). (c) The Iversen property.

If the set $\mathfrak{E}$ of points at which a map φ is not of type-$B\ell$ is not empty, then it is locally of positive capacity at each of its points (i.e. each neighborhood of each of its points contains a part of $\mathfrak{E}$ of positive capacity). This fact leads to the theorem: If $F \in O_{HB}$ (the class of Riemann surfaces admitting no non-constant bounded harmonic functions), then a conformal map with domain F is locally of type-$B\ell$. In particular, maps with parabolic domains are locally of type-$B\ell$. Prime instances of maps of type-$B\ell$ are the entire functions and the meromorphic functions in the finite plane. Simple examples show that we cannot expect the more refined results of the theory of meromorphic functions in the finite plane (such as the Picard theorem) to persist when we turn to domains of infinite genus even when such domains are parabolic Riemann surfaces with one ideal boundary element (e.g. $\sqrt{e^z + 1}$). However we have in the notion of a map which is locally of type-$B\ell$ a source of some positive information on the covering properties of a map whose domain may be quite complicated from a topological point of view.

As far as I am aware, no systematic study has been made hitherto of the covering properties of the "multiple-valued" transcendental functions such as abelian integrals and solutions (considered globally) of linear differential equations in the complex domain. The elliptic integrals of the first kind constitute a highly suggestive example, although their case is prejudiced by being pervaded with parabolicity. In the last section of the present essay we shall see that the abelian integrals and the solutions of linear differential equations of a fairly extensive class are meromorphic functions which are locally of type-$B\ell$ (relative to the extended plane) even though their domains may have a "strong" ideal boundary or may be of considerable complexity from a topological point of view.

5. The localization principle

While it would take us too far afield to elaborate the arguments on which the results of the preceding section are based, there is one link between the local and global aspects which should be brought out—*the localization principle.*

Suppose that U is a non-negative harmonic function on a Riemann surface F and that $\{\Omega_k\}$ is a family of regions of F satisfying: (1) $\Omega_k \cap \Omega_\ell = \emptyset$ for $k \neq \ell$, (2) for each k, for each $p \in fr\, \Omega_k$ there exists a continuum C satisfying $p \in C \subset fr\, \Omega_k$. Suppose further that U_k is a non-negative harmonic function on Ω_k, that U_k vanishes continuously on $fr\, \Omega_k$ and that $U \geq U_k$ on Ω_k. Let U_k^* denote the least harmonic majorant of the subharmonic function which is equal to U_k on Ω_k and to zero on $F - \Omega_k$. Then evidently

$$\Sigma_k U_k^* \leq U.$$

If U_k is quasi-bounded (singular) on Ω_k, then U_k^* is quasi-bounded (singular) on F. Further a (convergent) sum of quasi-bounded (singular) non-negative harmonic functions is quasi-bounded (singular). Hence with V_k and W_k denoting respectively the quasi-bounded and singular components of U_k, we see that ΣV_k^* and ΣW_k^* are respectively the quasi-bounded and singular components of ΣU_k^*.

Let us apply this observation to the localization of the residual term. We are considering a conformal map $\varphi : F \to G$ (hyperbolic), $q \in G$, Ω a simply-connected Jordan region in G containing q. Let $\{R_k\}$ denote the set of components of $\varphi^{-1}(\Omega)$, $R_k \neq R_\ell$ for $k \neq \ell$. Let $u_{k,q}$ denote the residual term associated with the map $\varphi_k : R_k \to \Omega$ where φ_k denotes the restriction of φ to R_k. Let $v_{k,q}$ and $w_{k,q}$ denote respectively the quasi-bounded and singular components of $u_{k,q}$. Since $u_{k,q}$ vanishes continuously on $fr\ R_k$ and $u_{k,q}(p) \leq \mathfrak{G}_G(\varphi(p), q)$, we infer that $u_{k,q}^* \leq u_q$ and hence that

$$\Sigma_k u_{k,q}^* \leq u_q.$$

On the other hand, with $A = \max_{r \in fr \Omega} \mathfrak{G}_G(r, q)$, we have $\mathfrak{G}_G(r, q) \leq \mathfrak{G}_\Omega(r, q) + A$, $r \in \Omega$ and $\mathfrak{G}_G(r, q) \leq A$, $r \in G - \Omega$, so that $u_q(p) \leq A$ for

$p \in \varphi^{-1}(\Omega)$ and $u_q(p) \leq A + \mathfrak{G}_\Omega(\varphi(p), q)$, $p \in R_k$. We have $u_q(p) \leq A + u_{k,q}(p)$, $p \in R_k$. Hence $u_q \leq A + \Sigma u^*_{k,q}$. This leads to *the localization principle*:

$$(5.1) \qquad \begin{cases} 0 \leq u_q - \Sigma_k u^*_{k,q} = O(1), \\ w_q = \Sigma_k w^*_{k,q}. \end{cases}$$

The first relation has as an immediate consequence the fact that a map of type-$B\ell$ is of type-$B\ell$ at each range point. The second relation reveals how the singular component behaves under localization. This result has important consequences, in particular it leads to a notion which does not appear to have been studied even in the classical case—the harmonic dimension of an asymptotic spot.

6.

It is to be recalled that a positive harmonic function u on a Riemann surface F is termed a *minimal* positive harmonic function (of F) provided that the positive harmonic functions on F which u dominates are proportional to u. This fundamental concept in the theory of positive harmonic functions is due to R. S. Martin [13]. From our point of view it is essential to note that, if U_k is minimal in Ω_k, then U^*_k is minimal in F. Further we note that, if a positive harmonic function admits a representation as a sum of mutually non-proportional minimals, then such a representation is unique.

1. Asymptotic spots. We return once again to an unrestricted conformal map $\varphi : F \to G$. Given $q \in \overline{\varphi(F)}$, we introduce the family Φ_q of simply-connected Jordan regions of G which contain q. By an *asymptotic spot of φ over q* is meant a function σ whose domain is Φ_q which satisfies (1) for each $\Omega \in \Phi_q$, $\sigma(\Omega)$ is a component of $\varphi^{-1}(\Omega)$ which is *not* relatively compact, (2) if $\Omega_1 \subset \Omega_2$, then $\sigma(\Omega_1) \subset \sigma(\Omega_2)$. When we say "$\sigma$ is an asymptotic spot" without further qualification, we mean that σ is an asymptotic spot of φ over some $q \in \overline{\varphi(F)}$ (q is necessarily unique).

Given an asymptotic spot σ of φ over q, we denote by $u_{\sigma(\Omega)}$ the residual term associated with q and $\varphi_{\sigma(\Omega)} : \sigma(\Omega) \to \Omega$, $\varphi_{\sigma(\Omega)}$ being the restriction of φ to $\sigma(\Omega)$. That is,

$$u_{\sigma(\Omega)} = \text{G.H.M. } \mathfrak{G}_\Omega(\varphi_{\sigma(\Omega)}, q).$$

We associate with the pair (σ, Ω) an index $H(\sigma, \Omega)$ as follows. If $u_{\sigma(\Omega)} = 0$. we set $H(\sigma, \Omega) = 0$. If $u_{\sigma(\Omega)} > 0$ and is representable as a finite sum of mutually non-proportional minimal positive harmonic functions on $\sigma(\Omega)$, in number n, we set $H(\sigma, \Omega) = n$. In the remaining case, we set $H(\sigma, \Omega) = +\infty$. It is to be observed that as a consequence of (5.1), $H(\sigma, \Omega)$ is monotone in Ω, i.e. if $\Omega_1 \subset \Omega_2$, then $H(\sigma, \Omega_1) \leq H(\sigma, \Omega_2)$.

We define the *harmonic index*, $H(\sigma)$, of σ as simply inf $H(\sigma, \Omega)$ where all Ω in the domain of σ are taken into account.

Thanks to the notion of harmonic index, it is possible to formulate the Denjoy-Carleman-Ahlfors theorem for meromorphic functions more precisely. In fact, let f denote a non-constant meromorphic function in the finite plane. Let $T(r; f)$ denote the Nevanlinna characteristic function of f. Let $H = \Sigma_\sigma H(\sigma)$. We have [9c]

THEOREM 4. *If* $H = +\infty$, *then* $\lim_{r\to+\infty} \log T(r; f)/\log r = +\infty$. *If* $2 \leq H < +\infty$, *then* $\liminf_{r\to\infty} r^{-H/2}T(r; f) > 0$. *If* $H = 1$ *and* σ_0, *the asymptotic spot with index one, is such that for some* Ω *of its domain, the complement of* $\sigma_0(\Omega)$ *intersects all circles* $\{|z| = r\}$ *with* r *sufficiently large, then* $\liminf_{r\to\infty} r^{-1/2}T(r; f) > 0$.

A weaker result of this type has been given by A. A. Gol'dberg [8] who established a theorem which is essentially equivalent to Theorem 4 with H replaced by $\Sigma_\sigma \min \{H(\sigma), 1\}$. The argument of Gol'dberg is a straightforward adaptation of the standard one given for the Denjoy-Carleman-Ahlfors theorem in the meromorphic case. Its essential contribution is to observe that the usual hypothesis involving direct transcendental singularities is unnecessarily stringent and to replace it by a condition which puts one immediately on the road of the classical argument. In order to arrive at the more refined version given by Theorem 4 account has to be taken of the harmonic structure of $u_{\sigma(\Omega)}$ for small Ω. It is through the use of the following decomposition theorem for positive harmonic functions and a device of Kjellberg [11] that one may bring into play the conventional Denjoy-Carleman-Ahlfors theorem argument in such a manner that the full force of H asserts itself.

Let u *be a positive harmonic function on a Riemann surface* F. *Let* n *denote a positive integer. If* u *is not representable as the sum of at most* n *mutually non-proportional minimals on* F, *then there exist* $n + 1$ *positive harmonic functions* $u_1, \ldots, u_{n+1}$ *on* F *satisfying:* $u = \Sigma u_k$; G.H.M. min $\{u_k, u_\ell\} = 0, k \neq \ell$.

The question immediately arises whether the harmonic index may assume finite values >1 and whether under such circumstances the lower growth estimate afforded by Theorem 4 is best possible. The function $z \cos (z^{n/2})$, n a positive integer, is an example of an entire function f satisfying the following conditions: (a) f has a sole asymptotic spot, say σ_0; (b) σ_0 lies over ∞ and $H(\sigma_0) = n$; (c) f is of order $n/2$. Thus we are assured of the existence of asymptotic spots having an assigned finite harmonic index >1. In the example just cited the growth is correctly estimated.

On the other hand, it is to be observed that in the case of a non-constant entire function of finite lower order, the *harmonic index of an asymptotic spot over a finite point cannot exceed one*. The source of this phenomenon is to be found in the fact that each component of the antecedent of a

Jordan region lying in the finite plane with respect to a non-constant entire function is simply-connected. (This property of the harmonic index leads to the result that a direct transcendental critical point of the inverse of a non-constant entire function of finite lower order which lies over a *finite* point is actually *logarithmic*. For such a function the result persists even more generally for asymptotic spots associated with finite locally omitted values [9c].)

These observations give rise to the following realization problem which appears as a natural extension of the theory of the Nevanlinna transcendental functions: Let $w_1, \ldots, w_n$ denote n (≥ 1) given points of the extended plane and let $H_1, \ldots, H_n$ denote n given positive integers. Does there exist a meromorphic function f in the finite plane satisfying: (a) the asymptotic spots of f having positive harmonic index are n in number, say $\sigma_1, \ldots, \sigma_n$, (b) σ_k lies over w_k and $H(\sigma_k) = H_k$, (c) f is of order $H/2$, where $H = \Sigma H_k$?

The function $z \cos (z^{H/2})$ yields a positive solution when $n = 1$.

For the case, $n = 2$, $H_1 = H_2 = 2$, w_1 and w_2 unrestricted, an affirmative answer may also be given. It is to be observed that, if $w_1 \neq w_2$, f must take all values infinitely often. A starting point for such a construction is the entire function $g(z) \equiv \cos (z^2)e^{-iz}$. We consider the component Ω_m of $\{|g| > m\ (>1)\}$ which lies in the upper half-plane. On fixing m suitably and augmenting Ω_m by the compact components of its complement, we are led, on composing g with a suitably normalized univalent conformal map of the upper half-plane onto the augmentation of Ω_m, (thanks to the Schwarz reflection principle) to a meromorphic function in the finite plane which is of constant modulus on the real axis. It is this meromorphic function which furnishes a solution to our problem when $w_1 = 0$ and $w_2 = \infty$. The situation where w_1 and w_2 are unrestricted may be reduced to this normalized one.

The general problem which we have proposed appears to be fairly difficult. Its resolution would cast considerable light on the possible complexity of meromorphic functions as it is reflected in their local covering properties.

7. Lindelöfian maps

We note that a conformal map $\varphi : F \to G$ with G hyperbolic satisfies

$$S(p, q) \equiv \Sigma_{\varphi(r)=q} n(r;\ \varphi) \mathfrak{G}_F(p, r) < +\infty \tag{7.1}$$

for $\varphi(p) \neq q$. This is a trivial consequence of $S(p, q) \leq \mathfrak{G}_G(\varphi(p), q)$. Now it is natural to inquire into the properties of a conformal map $\varphi : F \to G$ where F is hyperbolic, but G need not be, and (7.1) is satisfied. We term such a map a *Lindelöfian map* because of its intimate connection with the Lindelöf principle. Of course, it is in the case of parabolic G that the study of Lindelöfian maps is of particular interest. What are the essential facts concerning Lindelöfian maps? A guide to their investigation is the special case: $F = \{|z| < 1\}$, $G =$ extended plane. Here the Lindelöfian maps turn out to be precisely the meromorphic functions with bounded Nevanlinna characteristic function.

Now it is the case that the notion of a Nevanlinna characteristic function may be introduced for an arbitrary conformal map $\varphi : F \to G$, and indeed in such a way that the Nevanlinna first fundamental theorem and its essential consequences persist.† In particular, the theorem of Nevanlinna and Frostman which relates the characteristic function with the behavior of the Nevanlinna "Anzahlfunktion" on a set of positive capacity remains valid. When account is taken of the fact that the counterpart of the Nevanlinna "Anzahlfunktion" in our generalized theory is simply $S(p_0, q)$ for φ_ω where φ_ω is the restriction of φ to a relatively compact region ω containing a given point p_0 (when $\varphi(p_0) \neq q$ and an obvious variant when $\varphi(p_0) = q$), it is natural that the Lindelöfian maps turn out to be those with bounded characteristic function.

With the aid of the extended first fundamental theorem of Nevanlinna we may establish the following propositions, of which the first is a necessary and the second is a sufficient condition for a map to be Lindelöfian. In what follows U_q will denote a function with domain G for which there exists a simply-connected Jordan region Ω of G containing q such that the restriction of U_q to Ω is the Green's function for Ω with pole at q and the restriction of U_q to $G - \Omega$ is identically zero.

A. *Given a Lindelöfian map $\varphi : F \to G$ and $q \in G$. Then the sub-harmonic function $p \to U_q(\varphi(p)) - S(p, q)$ possesses a least harmonic majorant, say u. Further $u \geq 0$ and the singular component of u is independent of admitted U_q.*

B. *If there exists $q_0 \in G$ such that $S(p, q_0) < +\infty$ for $\varphi(p) \neq q_0$ and $p \to U_{q_0}(\varphi(p)) - S(p, q_0)$ possesses a harmonic majorant, then φ is Lindelöfian.*

Among the important properties enjoyed by functions of bounded characteristic in the interior of the unit circle is the extension of Frostman and Nevanlinna of the celebrated theorem of F. and M. Riesz concerning the radial limits of bounded analytic functions. The problem arises whether the Riesz-Frostman-Nevanlinna theorem has its counterpart for Lindelöfian maps.

Let us first investigate a more broadly conceived problem. We consider a conformal map $\varphi : F \to G$ where F is hyperbolic but otherwise no

† This generalization of the Nevanlinna characteristic function was made independently by Parreau [17b] and myself [9b].

restrictions are imposed and a set $\mathfrak{E} \subset G$ and seek to measure in some sense the "affinity" that φ has for $\mathfrak{E}$ in the neighborhood of the ideal boundary of F. The preceding sentence is to be taken only as an expression of intent. The means of measuring that we propose is the function $B_{\mathfrak{E}}$ which we now define.

For each open set $O \subset G$, $\varphi^{-1}(O)$ is an open set of F. There exists a least non-negative superharmonic function P_O on F which dominates 1 on $\varphi^{-1}(O)$. Let $b_O =$ G.H.M. P_O. Let $B_{\mathfrak{E}}$ denote the lower envelope of the family $\{b_O\}_{O \supset \mathfrak{E}}$. We have

THEOREM 5. *$B_{\mathfrak{E}}$ is non-negative harmonic on F. It is monotone and finitely subadditive in $\mathfrak{E}$.*

If φ is Lindelöfian, then Cap $\mathfrak{E} = 0$ *implies* $B_{\mathfrak{E}} = 0$.

The Riesz-Frostman-Nevanlinna theorem is a corollary of the last assertion of this theorem. We shall want for later applications some information about null ideal boundaries from the present point of view. Theorem 6 is the mate here of the last sentence of Theorem 5.

THEOREM 6. *If G is parabolic and φ is Lindelöfian, then the lower envelope of $\{B_{G-C}\}$ (C compact) is zero.*

Simple examples such as

$$f(z) \equiv \frac{z + \alpha}{z(1 + \alpha z)}, \qquad |z| < 1,$$

where $0 < \alpha < 1$, suggests that it is reasonable to expect that the following theorem holds.

THEOREM 7. *If G is parabolic (possibly compact) and $\varphi : F \to G$ is Lindelöfian, then the set of points of G at which φ is of type-$B\ell$ is a proper subset of G.*

This theorem may be established as follows. Suppose that φ were of type-$B\ell$ at each point of G. Let $p_0 \in F$ and let Ω_0 denote a simply-connected Jordan region of G containing $\varphi(p_0)$. There would exist q_0 $(\neq \varphi(p_0)) \in \Omega_0$ such that G.H.M. $\mathfrak{G}_{\Omega_0}(\varphi_R, q_0) = 0$, where $R = \varphi^{-1}(\Omega_0)$. This is a consequence of: (1) the restriction of φ to a component C of R would be a map of type-$B\ell$ of C into Ω_0, and (2) the set of q in Ω_0 for which G.H.M. $\mathfrak{G}_{\Omega_0}(\varphi_C, q) > 0$ would be of zero capacity. As a consequence of the localization principle (5.1), for each relatively compact Jordan region Ω of G containing $\bar{\Omega}_0$ we would have, with ω denoting the component of $\varphi^{-1}(\Omega)$ containing p_0:

$$\mathfrak{G}_\Omega(\varphi(p_0), q_0) = \Sigma_{\varphi_\omega(r)=q_0} n(r;\ \varphi)\mathfrak{G}_\omega(p_0, r) \leq S(p_0, q_0).$$

But then $\sup_{\Omega \supset \Omega_0} \mathfrak{G}_\Omega(\varphi(p_0), q_0) < +\infty$ and the parabolicity of G would be violated.

8. Lindelöfian meromorphic functions

In the case where $G =$ extended plane, the Lindelöfian maps are meromorphic functions and are appropriately termed *Lindelöfian meromorphic functions.* We shall not examine them in extenso, but we do wish to point out a few interesting phenomena which occur in their study. They enjoy an algebraic cohesiveness. *The Lindelöfian meromorphic functions (together with the constants) on a given hyperbolic Riemann surface constitute a field.* However, in marked contrast to the classical situation for functions of bounded characteristic in the interior of the unit circle, there are Riemann surfaces which admit non-constant bounded analytic functions but are such that not all Lindelöfian functions on them admit representation as a quotient of bounded analytic functions. There exist also Riemann surfaces of the class O_{AB} which admit (non-constant) Lindelöfian meromorphic functions. Instances of such a surface are afforded by a plane region $\in O_{AB}$ which possesses a Green's function and by the celebrated example of P. J. Myrberg [15] of a 2-sheeted transcendental hyperelliptic surface less a simple disk—an example which is the source and impetus of many important investigations in the infinite genus theory.

9. A deformation problem

A Riemann surface F will be said to belong to the class $\mathfrak{D}$ provided that there exists a continuous map φ of $F \times [0, 1]$ into F such that (1) $\varphi_t : p \to \varphi(p, t)$ is conformal for each $t \in [0, 1]$, (2) $\varphi_0 \neq \varphi_1$. The notation "$\mathfrak{D}$" is used to suggest the fact that F admits a deformable conformal map into itself. Now if the domain of the universal covering of F is not hyperbolic, then it is easy to show that $F \in \mathfrak{D}$. On the other hand if F has a null boundary, but the domain of its universal covering is hyperbolic, then $F \notin \mathfrak{D}$ [9d].

There remains the case where F is hyperbolic. It is trivial that, if $F \notin O_{AB}$, then $F \in \mathfrak{D}$. Now it is actually the case that the converse holds. A detailed discussion will be given elsewhere in this Conference. The object of the present remarks is to show the decisive role played by Theorem 7.

A first step may be made by noting that F admits a meromorphic function, f, satisfying

$$\log |f| = \mathfrak{G}_F(\varphi_1, q) - \mathfrak{G}_F(\varphi_0, q),$$

q being so chosen that $f \neq$ const. Thanks to II of section 7, f is Lindelöfian, since $S(p, \infty) \leq \Sigma_{\varphi_1(r)=q} n(r;\ \varphi_1)\mathfrak{G}_F(p, r) \leq \mathfrak{G}_F(\varphi_1(p), q)$ and $\log^+ |f(p)| - S(p, \infty) \leq \mathfrak{G}_F(\varphi_1(p), q)$. By Theorem 7 we are assured of an abundance of non-constant bounded harmonic functions on F. The proof is reduced to showing that we may choose a non-constant bounded harmonic function b on F in such a manner that $b \circ \varphi_1 - b \circ \varphi_0$, which is the real part of an analytic function on F and is bounded, is not constant.

10. Certain classes of "multiple-valued" functions

Let φ denote a conformal covering of a Riemann surface G with domain a Riemann surface F. Let f denote a non-constant meromorphic function with domain F. We shall say that f is *φ-regular* provided that for each $q \in G$ there exists a simply-connected region Ω, $q \in \Omega \subset G$, such that the family $\{f \circ \psi_k\}$ is quasinormal† in Ω, the ψ_k being the inverses of the restrictions of φ to the components of $\varphi^{-1}(\Omega)$. It is readily seen that the set of *values* of the *constant* limits of an admitted family $\{f \circ \psi_k\}$ is independent of the family. We denote the set of values in question by C. We shall be concerned with the covering properties of f under the assumption that C is sufficiently sparse and G is parabolic. The following theorem is central here.

THEOREM 8. *If G is parabolic and* cap $C = 0$, *then f is locally of type-$B\ell$ in the extended plane.*

PROOF. Let α denote an asymptotic value of f which is not in C and let P denote an asymptotic path‡ on F satisfying $\lim_{t\to 1} f \circ P = \alpha$. We consider the behavior of $Q = \varphi \circ P$. We assert that *$Q(t)$ tends to the ideal boundary of G as $t \to 1$.* To see this, we first note that $Q(t)$ cannot tend to a point of G as $t \to 1$ since φ is a covering. If $Q(t)$ did not tend to the ideal boundary of G, there would exist: $q_0 \in G$, a simply-connected region Ω containing q_0, and increasing sequences of positive numbers (s_k) and (t_k) such that (1) $s_k < t_k < s_{k+1}$, $Q(s_k) \in \Omega$, and $Q(t_k) \in G - \Omega$ for all k; $\lim s_k = \lim t_k = 1$; $\lim Q(s_k) = q_0$; (2) the family $\{f \circ \psi_k\}$ associated with Ω is quasinormal. But these conditions imply $\alpha \in C$ and this is impossible.

We now let $\mathfrak{B}$ denote the set of points at which f is of type-$B\ell$. Clearly $\mathfrak{B} - C \subset \overline{f(F)} - C$. Once we show that $\overline{f(F)} - C \subset \mathfrak{B} - C$, noting that the complement of C with respect to the extended plane is connected and that $\overline{f(F)} - C \neq \emptyset$, we are assured that f is of type-$B\ell$ at each point

† We recall that this condition means that for each sequence (g_n) whose elements belong to $\{f \circ \psi_k\}$ there exists a subsequence (h_n) which converges uniformly in Ω less a finite set of points (the number depending on (h_n)). cf. Montel: *Familles Normales.*

‡ We recall that an asymptotic path of f is a continuous map P of $(0 \leq t < 1)$ into F such that (1) for each compact $K \subset F$, $P(t) \in F - K$ for t sufficiently near one, (2) $\lim_{t\to 1} f \circ P$ exists. To say that α is an asymptotic value of f is to assert the existence of an asymptotic path P such that $\lim f \circ P = \alpha$; this is equivalent to asserting the existence of an asymptotic spot of f over α.

of the complement of C. Now the set of points at which a conformal map is not of type-$B\ell$ has positive capacity if it is not empty [§4]. Since cap $C = 0$, we infer that $\mathfrak{B}$ is the extended plane.

We turn our attention to showing that $\overline{f(F)} - C \subset \mathfrak{B} - C$. Suppose that w is a point of $\overline{f(F)} - C$ at which f is not of type-$B\ell$. For each $r > 0$, there exists a component of $f^{-1}(\Delta_r)$, where $\Delta_r = \{|z - w| < r\}$ if $w \neq \infty$ and $\Delta_r = \{|z| > r^{-1}\}$ if $w = \infty$, such that the restriction of f to the component in question, $\mathfrak{C}_r$, is not a map of type-$B\ell$ of $\mathfrak{C}_r$ into Δ_r. Let h_r denote a conformal covering of $\mathfrak{C}_r$ with domain $\{|z| < 1\}$. Let $g_r = r^{-1}(f \circ h_r - w)$ if $w \neq \infty$ and let $g_r = r^{-1}(f \circ h_r)^{-1}$ if $w = \infty$. Then g_r is an analytic function of modulus less than one which is not of type-$B\ell$ relative to the unit disk. Hence the Fatou radial limits of g_r are of modulus less than one on a set of positive measure. It follows that $\varphi(h_r(\rho e^{i\theta}))$ tends to the ideal boundary of G for a set of θ, say Θ_r, of positive measure when r is sufficiently small.

This implies that the *image set of* $\varphi \circ h_r$ *is dense in* G for r small. Otherwise $\varphi \circ h_r$ would be a Lindelöfian map of the unit disk into G. But then by Theorem 6, Θ_r would have zero Lebesgue measure.

We fix a point $q \in G$ and a simply-connected Jordan region Ω containing q for which the associated family $\{f \circ \psi_k\}$ is quasinormal in Ω. For r sufficiently small a segment σ_r may be chosen in the unit disk which is such that its endpoints, a_r and b_r, are carried respectively into points of Ω and $G - \Omega$ and $\lim_{r \to 0} \varphi(h_r(a_r)) = q$. By paraphrasing the argument of the first paragraph of the present proof, we see that $w \in C$. This is impossible. We conclude that $\overline{f(F)} - C \subset \mathfrak{B}$ and the theorem follows.

In the special case where G is *compact* a more precise conclusion may be drawn. For here f does not have any asymptotic value in the complement of C. It follows by Theorem 14.1 of [9a] that *f is of type-$B\ell_1$ at each point of the complement of C.*

It is easy to construct an example where the situation of Theorem 8 is realized with C an *assigned* closed set of zero capacity and F is the unit disk. In fact, in case C contains at least three points, it suffices to take f as a conformal universal covering of the complement of C with respect to the extended plane, f having domain $\{|z| < 1\}$, and φ as a conformal universal covering of a compact Riemann surface G of genus ≥ 2. The case where C contains precisely one or two points is readily obtained from the case just treated on taking C as the set of the fourth roots of unity and taking f^2 or f^4 to replace f. The case where C is empty is essentially trivial. It suffices to take $f = g \circ \varphi$ where g is a non-constant meromorphic function on G.

1. φ-FINITE DIMENSIONAL FUNCTIONS. We return to a conformal covering φ of a Riemann surface G with domain a Riemann surface F and a

non-constant meromorphic function f also with domain F. For each simply-connected region $\Omega \subset G$ we consider the linear vector space $\mathfrak{V}_\Omega$ over the complex field generated out of $\{f \circ \psi_k\}$ where the ψ_k are again the inverses of the restrictions of φ to the components of $\varphi^{-1}(\Omega)$. Now dim $\mathfrak{V}_\Omega$ is *independent* of Ω. We term f a *φ-finite dimensional function* (of dimension d) provided that the common value, d, of dim $\mathfrak{V}_\Omega$ is finite. The values of the constant functions in $\mathfrak{V}_\Omega$ are independent of Ω. We term a φ-finite dimensional function *non-degenerate* provided that the only constant function in a $\mathfrak{V}_\Omega$ is 0.

The notion of a φ-finite dimensional function goes back to the celebrated paper [19] of Riemann on the hypergeometric function. The notion embraces as special instances multiplicative meromorphic functions and solutions of linear differential equations with meromorphic coefficients in a plane region.

We shall now see that *a non-degenerate φ-finite dimensional function is φ-regular and $C \subset \{0, \infty\}$*. It remains open whether the requirement of non-degeneracy is really essential to draw the conclusion of Theorem 8. At all events, the multiplicative functions and solutions of homogeneous differential equations for which the coefficient of the term of order zero does not vanish belong to the class of the non-degenerate φ-finite dimensional functions.

Suppose then that f is a non-degenerate φ-finite dimensional function. Let $q \in G$ and let Ω denote a simply-connected region of G containing q. Let $g_1, \ldots, g_d$ constitute a basis for $\mathfrak{V}_\Omega$. Let Δ denote a Jordan disk containing q whose closure is contained in Ω and is such that each g_k is analytic on $\bar{\Delta} - \{q\}$. We test the quasi-normality of $\{f \circ \psi_k\}$ in Δ. Given a sequence (h_n) whose members belong to $\{f \circ \psi_k\}$, for each n, we have

$$h_n = \sum_{k=1}^{d} \alpha_{nk} g_k.$$

If $\sup_n \left\{ \sum_{k=1}^{d} |\alpha_{nk}| \right\} < +\infty$, then it is immediate that there exists a subsequence of (h_n) which converges uniformly in $\Delta - \{q\}$. If $\sup_n \left\{ \sum_{k=1}^{n} |\alpha_{nk}| \right\} = +\infty$, there exists a subsequence of (h_n) which converges uniformly to ∞ in Δ less a finite set. The quasi-normality is established.

What are the possible values of the constant limit functions? Apart from ∞, the only possibility is 0 because of the non-degeneracy hypothesis. Hence $C \subset \{0, \infty\}$.

A study of the hypergeometric differential equation associated with the modular function leads to an example of an analytische Gebilde $\mathfrak{A}$ with the following properties: (1) the center function, c, is a universal covering of the plane punctured in 0, 1, ∞, (2) the elements of $\mathfrak{A}$ annihilate the hypergeometric differential equation in question, (3) the value function is non-degenerate c-finite dimensional, the dimension being 2. Thus we see that in the classical theory of linear differential equations with a finite number of singular points we must be prepared to cope with solutions whose analytische Gebilde have "very strong boundaries."

We remark that even in the study of multiplicative functions one may encounter coverings of some complexity, in fact coverings having domains of infinite genus.

2. ABELIAN DIFFERENTIALS. Let ω denote a meromorphic differential of the first order (different from the zero differential) on a parabolic Riemann surface G. We shall denote the order of ω at q by $O(q;\ \omega)$. Let f denote a primitive of ω and let F denote the domain of f. There is associated with f a conformal covering φ of $G - \{q \mid O(q;\ \omega) < 0\}$, whose domain, R, is F less the poles of f. Now f_R, the restriction of f to R, is φ-regular. The set C contains at most ∞. We conclude from Theorem 8 that f_R is locally of type-$B\ell$. It follows that f is also locally of type-$B\ell$.

3. ABELIAN DIFFERENTIALS OF THE FIRST KIND; *G compact.* If G is compact and has positive genus and if ω is of the first kind, then f has no finite asymptotic values. It is possible to make even more precise statements concerning the covering properties of f in this case. Let $N = \max_{q \in G} O(q;\ \omega)$. We shall show:

There exists $\mu > 0$ such that for each finite w each component $\mathfrak{C}$ of $f^{-1}(\Delta_\mu(w))$, where $\Delta_\mu(w) = \{|z - w| < \mu\}$, is simply-connected and relatively compact and the restriction of f to $\mathfrak{C}$ is a map of constant valence ($\leq N + 1$) onto $\Delta_\mu(w)$ which has precisely one ramification point in $\mathfrak{C}$.

PROOF. For each $q \in G$ there exists a simply-connected region Ω_q of G containing q such that each member of the associated family $\{f \circ \psi_k\}$ (ψ_k as above relative to Ω_q) is a map of finite constant valence ($= O(q;\ \omega) + 1$) of Ω_q onto a plane circular disk. The radius of this disk is independent of k and will be denoted by ρ_q. Let g_q denote a member of the family $\{f \circ \psi_k\}$. We let V_q denote a simply-connected Jordan region containing q whose closure lies in Ω_q. There exists a finite subset of $\{V_q\}$ covering G, say $V_{q_1}, \ldots, V_{q_n}$. We take μ satisfying

$$0 < 2\mu < \min \{\rho_{q_k} - \max_{r \in \bar{V}_{q_k}} |g_{q_k}(r) - g_{q_k}(q_k)|\}. \tag{9.1}$$

Suppose that $w \in f(F)$. Let $\mathfrak{C}$ denote a component of $f^{-1}(\Delta_\mu(w))$ and fix $p \in \mathfrak{C}$. Suppose that $\varphi(p) \in V_{q_k}$. On considering $f \circ \psi_l$ associated with Ω_{q_k} where $\psi_l[\varphi(p)] = p$, we see by (9.1) that $\mathfrak{C}$ is relatively compact and that the restriction of f to $\mathfrak{C}$ is a map of constant valence ($\leq N + 1$) of $\mathfrak{C}$ onto $\Delta_\mu(w)$ which is ramified at most at one point. Further $f(F)$ is the finite plane since $w \in f(F)$ implies $\Delta_\mu(w) \subset f(F)$. (This may also be concluded by noting the absence of finite asymptotic values.)

In the special case where the genus of G is one, $O(q;\ \omega) = 0$, $q \in G$. The theorem which has just been established yields another proof of the classical result that in the genus one case f is a univalent map of F onto the finite plane. For f is a covering of the finite plane.

The developments of the present section represent only a beginning of the study of some of the important "multiple-valued" transcendental functions. The whole question appears to be worthy of further study.

REFERENCES

[1] LARS AHLFORS, Zur Theorie der Ueberlagerungsflächen, *Acta Math. 65* (1935), pp. 157–194.

[2] HEINRICH BEHNKE und KARL STEIN, Entwicklung analytischen Funktionen auf Riemannschen Flächen, *Math. Ann. 120* (1949), pp. 430–461.

[3] LIPMAN BERS, On rings of analytic functions, *Bull. A.M.S. 54* (1948), pp. 311–315.

[4] SALOMON BOCHNER, Fortsetzung Riemannscher Flächen, *Math. Ann. 98* (1927), pp. 406–421.

[5] HERTA FLORACK, Reguläre und meromorphe Funktionen auf nicht geschlossenen Riemannschen Flächen, *Schr. Math. Inst. Univ. Münster, No. 1* (1948), 34 pp.

[6] LÉONCE FOURÉS, (a) Sur la théorie des surfaces de Riemann, *Ann. Sci. École Norm. Sup. (3) 68* (1951), pp. 1–64. (b) Recouvrements de surfaces de Riemann, *Ann. Sci. École Norm. Sup. (3) 69* (1952), pp. 183–201.

[7] OTTO FROSTMAN, Potential d'équilibre et capacité des ensembles. Thesis. Lund, 1935.

[8] A. A. GOL'DBERG, On the influence of the clustering of algebraic branch points of a Riemann surface on the order of growth of a meromorphic mapping function (Russian), *Dokl. Akad. Nauk SSR (N.S.) 98* (1954), pp. 709–711; correction *101* (1955), p. 4.

[9] MAURICE HEINS, (a) On the Lindelöf principle, *Ann. of Math. 55* (1955), pp. 296–317. (b) Lindelöfian maps, *Ann. of Math. 62* (1955), pp. 418–445. (c) Asymptotic spots of entire and meromorphic functions, *Ann. of Math. 66* (1957), pp. 430–439. (d) A theorem concerning the existence of deformable conformal maps, *Ann. of Math. 67* (1958), pp. 42–44. (e) Algebraic structure and conformal mapping, *Trans. A.M.S. 89* (1958), pp. 267–276. (f) Meromorphic functions with assigned asymptotic values, *Duke Math. J. 22* (1955), pp. 353–356.

[10] SHIZUO KUKUTANI, Ring of analytic functions, *Lectures on Functions of a Complex Variable*, Univ. of Michigan (1955), pp. 71–83.

[11] BO KJELLBERG, On the growth of minimal positive harmonic functions in a plane region, *Ark. Mat. 1* (1950), pp. 347–351.

[12] OLLI LEHTO, Value distribution and boundary behavior of a function of bounded characteristic and the Riemann surface of its inverse function, *Ann. Acad. Sci. Fenn., Series A, No. 177* (1954).

[13] ROBERT S. MARTIN, Minimal positive harmonic functions, *Trans. A.M.S. 49* (1941), pp. 137–172.

[14] MARSTON MORSE, Topological methods in the theory of functions of a complex variable, *Ann. of Math. Studies 15* (1947).

[15] P. J. MYRBERG, Ueber die analytische Fortsetzung von beschränkten Funktionen, *Ann. Acad. Sci. Fenn., Series A, No. 1, Math. Phys. No.* 58 (1949), 7 pp.

[16] MAKOTO OHTSUKA, On a covering surface over an abstract Riemann surface, *Nagoya Math. J. 4* (1952), pp. 109–118.

[17] MICHEL PARREAU, (a) Sur les moyennes des fonctions harmoniques et analytiques et la classification des surfaces de Riemann, *Ann. Inst. Fourier Grenoble* (1951), pp. 103–197. (b) Fonction caractéristique d'une application conforme, *Ann. Fac. Sci. U. Toulouse* (1956), pp. 175–189.

[18] ALBERT PFLUGER, *Theorie der Riemannschen Flächen*, Berlin–Göttingen–Heidelberg, 1957.

[19] BERNHARD RIEMANN, *Gesammelte Werke*. 2 Aufl. Leipzig, 1892.

[20] HALSEY L. ROYDEN, Rings of analytic and meromorphic functions, *Trans. A.M.S. 83* (1956), pp. 269–276.

[21] WALTER RUDIN, An algebraic characterization of conformal equivalence, *Bull. A.M.S. 61* (1955), p. 543 (Abstract).

[22] STANISŁAW SAKS and ANTONI ZYGMUND, *Analytic functions*, Warsaw–Wrocław, 1952.

[23] HERMANN WEYL, *Die Idee der Riemannschen Fläche*, 3 Aufl. Stuttgart, 1955.

[24] GORDON T. WHYBURN, *Analytic topology*, New York, 1942.

On Certain Coefficients of Univalent Functions†

James A. Jenkins
NOTRE DAME UNIVERSITY

1. Despite the passage of years, Teichmüller's coefficient theorem [15] remains the most penetrating explicit result known in the general coefficient problem for univalent functions. It is strange, then, that these considerations have up to the present been of little use in finding explicit numerical bounds for the coefficients of lower order. The principal purpose of the present account is to show how a closely related method proves exceedingly effective in finding such bounds.

2. If $\mathfrak{R}$ is a Riemann surface, by a quadratic differential defined on $\mathfrak{R}$ we mean an entity which assigns to every local uniformizing parameter z of $\mathfrak{R}$ a function $Q(z)$ meromorphic in the neighborhood for z and satisfying the following condition. If z^* is a second local uniformizing parameter of $\mathfrak{R}$ and $Q^*(z^*)$ is the corresponding function associated with z^* and if the neighborhood on $\mathfrak{R}$ for z^* overlaps that for z, then at common points of these neighborhoods we have

$$Q^*(z^*) = Q(z)\left(\frac{dz}{dz^*}\right)^2.$$

Quadratic differentials will be denoted generically by symbols such as $Q(z)dz^2$. It is clear that we may speak of a quadratic differential having zeros and poles of a certain order. Zeros and simple poles of a quadratic differential are called finite critical points and the totality of them will be denoted generically by C. The totality of poles of order at least two will be denoted generically by H. A maximal regular curve on which $Q(z)dz^2 > 0$ is called a trajectory of $Q(z)dz^2$. A maximal regular curve on which $Q(z)dz^2 < 0$ is called an orthogonal trajectory of $Q(z)dz^2$. These curves are evidently independent of the choice of local uniformizing parameters.

In the neighborhood of a point of $\mathfrak{R}$ which is neither a zero nor a pole

† Research supported in part by the Mathematics Section, National Science Foundation.

of $Q(z)dz^2$ the trajectories behave like a regular curve family. A zero of order μ or a simple pole ($\mu = -1$) is the limiting end point of $\mu + 2$ trajectory arcs equally spaced at angles of $2\pi/(\mu + 2)$. At a pole of order 2 the trajectories behave locally either like radial arcs, logarithmic spirals, or concentric circles. At a pole of order $\mu(\mu > 2)$ there are $\mu - 2$ asymptotic directions for trajectories equally spaced at angles of $2\pi/(\mu - 2)$. More detailed accounts of the local structure of trajectories will be found in [13, 11, 8, 10].

Teichmüller enunciated the principle that the solution of a certain type of extremal problem in Geometric Function Theory is in general associated with a quadratic differential. If in the problem a point is assumed to be fixed without further requirement the quadratic differential will have a simple pole there. If in addition the functions treated in the problem are required to have at the point, in terms of suitably assigned local uniformizing parameters, fixed values for their first n derivatives, the quadratic differential will have a pole of order $n + 1$ there. More generally, the highest derivative occurring may not be required to be fixed but some condition on its region of variation may be desired. Teichmüller was led to this principle by abstraction from the numerous results of Grötzsch and by his own considerations on quasiconformal mappings. However he never gave anything in the nature of an explicit general result embodying this principle, the most general being his coefficient theorem to be discussed below.

3. We will now give a statement of the General Coefficient Theorem which, while it does not exhaust the possibilities of application of Teichmüller's principle, does constitute a precise formulation of this principle for a wide class of extremal problems and does include as special cases almost all standard results for univalent functions. We confine our attention to finite oriented Riemann surfaces. On such a surface a quadratic differential is understood to be real in terms of boundary uniformizing parameters and to be regular at boundary points apart from possible simple poles.

Let $\mathfrak{R}$ be a finite oriented Riemann surface, $Q(z)dz^2$ a quadratic differential on $\mathfrak{R}$. By an admissible family $\{\Delta\}$ of domains $\Delta_j, j = 1, \ldots, k$, on $\mathfrak{R}$ with respect to $Q(z)dz^2$ we mean the complement on $\mathfrak{R}$ of the union of a finite set of trajectories of $Q(z)dz^2$ each of which is either closed or has a limiting end point in each sense at a point of C, possible end points of these trajectories and a finite number of arcs in $\mathfrak{R} - H$ on closures of trajectories.

Let $\{\Delta\}$ be an admissible family of domains Δ_j, $j = 1, \ldots, k$, on the finite oriented Riemann surface $\mathfrak{R}$ with respect to the quadratic differential $Q(z)dz^2$. Let $\{f\}$ be a family of functions $f_j, j = 1, \ldots, k$, with the following properties:

(i) f_j maps Δ_j conformally into $\mathfrak{R}$,

(ii) if a pole A of $Q(z)dz^2$ lies in $\Delta_j, f_j(A) = A$,

(iii) $f_j(\Delta_j) \cap f_\ell(\Delta_\ell) = 0, j \neq \ell, j, \ell = 1, \ldots, k$.

Then the family $\{f\}$ is said to admit an admissible homotopy F into the identity if there exists a function $F(P, t)$ defined for $P \in \bigcup_{j=1}^{k} \Delta_j, 0 \leq t \leq 1$, with values in $\mathfrak{R}$, continuous in both variables together, satisfying the following conditions:

(a) $F(P, 0) = f_j(P), P \in \Delta_j, j = 1, \ldots, k$,

(b) $F(P, 1) = P, P \in \bigcup_{j=1}^{k} \Delta_j$,

(c) $F(P, t) = P$, P a pole of $Q(z)dz^2$ in $\bigcup_{j=1}^{k} \Delta_j, 0 \leq t \leq 1$,

(d) $F(P, t) \neq Q$, Q a pole of $Q(z)dz^2$ in $\mathfrak{R}$, $P \neq Q, 0 \leq t \leq 1$.

Let $\{f\}$ be a family of functions satisfying the preceding conditions (i), (ii), (iii) and admitting an admissible homotopy F into the identity. Let A be an element of H and N the neighborhood of a local uniformizing parameter z in terms of which A is represented by $z = 0$. For P sufficiently near to A, $F(P, t), 0 \leq t \leq 1$, will lie in $N - \{A\}$. Let $\delta(F, P)$ denote the change in argument as we describe the image of $F(P, t)$ from the image of $F(P, 1)$ to that of $F(P, 0)$ in the z-plane. Then it is readily seen that

$$\lim_{P \to A} \delta(F, P)$$

exists and is independent of the choice of local uniformizing parameter z. This limit will be called the deformation degree of the family $\{f\}$ for the homotopy F at A and is denoted by $d(F, A)$.

Let $\{\Delta\}$ be an admissible family of domains $\Delta_j, j = 1, \ldots, k$, on the finite oriented Riemann surface $\mathfrak{R}$ with respect to the quadratic differential $Q(z)dz^2$. Then by an admissible family $\{f\}$ of functions $f_j, j = 1, \ldots, k$, associated with $\{\Delta\}$ we mean a family with the following properties:

(i) f_j maps Δ_j conformally into $\mathfrak{R}, j = 1, \ldots, k$,

(ii) if a pole A of $Q(z)dz^2$ lies in $\Delta_j, f_j(A) = A$,

(iii) $f_j(\Delta_j) \cap f_\ell(\Delta_\ell) = 0, j \neq \ell, j, \ell = 1, \ldots, k$,

(iv) if A is a pole of order m greater than two of $Q(z)dz^2$ in Δ_j in terms of a local parameter z representing A as the point at infinity $f_j(z)$ admits locally the representation

$$f_j(z) = z + \frac{a_{m-3}}{z^{m-3}} + \text{higher powers of } z^{-1},$$

(v) the family $\{f\}$ admits an admissible homotopy F into the identity,

(vi) the homotopy F can be chosen so that if A is a pole of order m greater than two of $Q(z)dz^2$ on the boundary of a strip domain [10, Ch. III] then

$$d(F, A) = 0.$$

In (iv) it is understood that $f_j(z)$ and z are both evaluated in terms of the same local parameter.

We are now ready to state the

GENERAL COEFFICIENT THEOREM. *Let $\mathfrak{R}$ be a finite oriented Riemann surface, $Q(z)dz^2$ a positive quadratic differential on $\mathfrak{R}$, $\{\Delta\}$ an admissible family of domains $\Delta_j, j = 1, \ldots, k$, on $\mathfrak{R}$ relative to $Q(z)dz^2$ and $\{f\}$ an admissible family of functions $f_j, j = 1, \ldots, k$, associated with $\{\Delta\}$. Let $Q(z)dz^2$ have double poles $P_1, \ldots, P_r$ and poles $P_{r+1}, \ldots, P_n$ of order greater than two. We allow either of these sets to be void but not both. Let $P_j, j \leq r$, lie in domain Δ_ℓ and in terms of a local parameter z representing P_j as the point at infinity let f_ℓ have the expansion*

$$f_\ell(z) = a^{(j)}z + a_0^{(j)} + \frac{a_1^{(j)}}{z} + \text{higher powers of } z^{-1}$$

and Q the expansion

$$Q(z) = \frac{\alpha^{(j)}}{z^2} + \text{higher powers of } z^{-1}.$$

Let $P_j, j > r$, a pole of order m_j greater than two, lie in the domain Δ_ℓ and in terms of a local parameter z representing P_j as the point at infinity let f_ℓ have the expansion

$$f_\ell(z) = z + \frac{a_{m_j-3}^{(j)}}{z^{m_j-3}} + \text{higher powers of } z^{-1}$$

and Q the expansion

$$Q(z) = \alpha^{(j)}z^{m_j-4} + \text{decreasing powers of } z.$$

Then

$$\mathfrak{R}\left\{\sum_{j=1}^{r} \alpha^{(j)} \log a^{(j)} + \sum_{j=r+1}^{n} \alpha^{(j)} a_{m_j-3}^{(j)}\right\} \leq 0 \tag{1}$$

where $\log a^{(j)} = \log |a^{(j)}| - id(F, P_j), j \leq r$.

If equality occurs in (1) f_ℓ reduces to the identity in a domain Δ_ℓ for which any of the following conditions holds.

(i) *There is a pole of $Q(z)dz^2$ of order greater than two in Δ_ℓ.*

(ii) *There is a pole $P_j, j \leq r$, with the corresponding coefficient $a^{(j)}$ equal to one in Δ_ℓ.*

(iii) *There is a simple pole of $Q(z)dz^2$ or a point on a trajectory ending in a simple pole in Δ_ℓ.*

Equality can occur in (1) when there exists a double pole $P_j, j \leq r$, such *that for the corresponding coefficient* $|a^{(j)}| \neq 1$ *only when* $\mathfrak{R}$ *is conformally equivalent to the sphere and* $Q(z)dz^2$ *is a quadratic differential whose only critical points are two poles each of order two. If further there is a single domain in the family* $\{\Delta\}$ *the corresponding function is conformally equivalent to a linear transformation with the points corresponding to these poles as fixed points.*

A detailed proof of this result is found in [10, Ch. IV]. A more special case is treated in [9].

4. We are interested here in two families of univalent functions, first the family S of functions $f(z)$ regular and univalent for $|z| < 1$ with expansion at the origin

$$f(z) = z + A_2 z^2 + A_3 z^3 + A_4 z^4 + \dots, \tag{2}$$

second the family Σ of functions $f(z)$ univalent for $|z| > 1$, regular apart from a simple pole at the point at infinity and having expansion at that point

$$f(z) = z + c_0 + \frac{c_1}{z} + \frac{c_2}{z^2} + \frac{c_3}{z^3} + \dots. \tag{3}$$

One form of Teichmüller's coefficient theorem is the following.

Let the quadratic differential $Q(w)dw^2$ *be regular on the* w*-sphere apart from a pole of order* $n + 4$ *at the point at infinity* $(n \geq -1)$ *at which it has the expansion*

$$Q(w)dw^2 = (\alpha w^n + \text{decreasing powers of } w)dw^2.$$

Let D *be an admissible domain with respect to the quadratic differential* $Q(w)dw^2$ *which is the image of* $|z| > 1$ *under a mapping* $w = f^*(z)$ *where* $f^* \in \Sigma$ *and* f^* *has expansion at the point at infinity*

$$f^*(z) = z + b_0 + \frac{b_1}{z} + \dots + \frac{b_k}{z_k} + \dots.$$

Let $f \in \Sigma$ *and have expansion* (3) *at the point at infinity where*

$$c_j = b_j, \quad j = 0, 1, \dots, n.$$

Then

$$\mathfrak{R}\{\alpha c_{n+1}\} \leq \mathfrak{R}\{\alpha b_{n+1}\}$$

equality occurring only for $f(z) \equiv f^*(z)$.

Let $\phi(w)$ be the inverse of the function $f^*(z)$ defined on D. We will apply the General Coefficient Theorem with $\mathfrak{R}$ the w-sphere, the quadratic differential $Q(w)dw^2$ and the admissible domain D. The function $f(\phi(w))$ has expansion at the point at infinity

$$w + \frac{c_{n+1} - b_{n+1}}{w^{n+1}} + \text{higher powers of } w^{-1}.$$

To see that this function is an admissible function associated with D we observe first that conditions (i), (ii), (iii) are automatically satisfied and the above expansion shows that the same is true of condition (iv). Moreover in the present simple topological situation the existence of an admissible homotopy is immediate and it can evidently be chosen so that the corresponding deformation degree at the point at infinity is zero in any case.

The quadratic differential has a single pole P_1 of order $n + 4$ at the point at infinity. The corresponding coefficients are

$$\alpha^{(1)} = \alpha,\ a_{n+1}^{(1)} = c_{n+1} - b_{n+1}.$$

Application of inequality (1) then gives

$$\Re\{\alpha(c_{n+1} - b_{n+1})\} \leq 0$$

or

$$\Re\{\alpha c_{n+1}\} \leq \Re\{\alpha b_{n+1}\}.$$

The equality statement follows from equality condition (i) of the General Coefficient Theorem.

The other form of Teichmüller's coefficient theorem is as follows.

Let the quadratic differential $Q(w)dw^2$ be regular on the w-sphere apart from a pole of order $n + 1$ at the origin ($n \geq 2$) at which it has the expansion

$$Q(w)dw^2 = \left(\frac{\alpha}{w^{n+1}} + \text{decreasing powers of } w^{-1}\right)dw^2$$

and a possible simple pole at the point at infinity. Let D be an admissible domain with respect to the quadratic differential $Q(w)dw^2$ not containing the point at infinity which is the image of $|z| < 1$ under a mapping $w = f^*(z)$ where $f^* \in S$ and f^* has expansion at the origin

$$f^*(z) = z + A_2^* z^2 + \ldots + A_k^* z^k + \ldots .$$

Let $f \in S$ and have expansion (2) *at the origin where*

$$A_j = A_j^*, \quad j = 2, \ldots, n - 1.$$

Then

$$\Re\{\alpha A_n\} \geq \Re\{\alpha A_n^*\}$$

equality occurring only for $f(z) \equiv f^*(z)$.

This statement is derived from the General Coefficient Theorem in a manner similar to the preceding one.

The precise bounds known for coefficients of functions in S and Σ are (in the notation of the expansions (2) and (3)): $|A_2| \leq 2$ [1], $|A_3| \leq 3$ [12], $|A_4| \leq 4$ [3], $|c_1| \leq 1$ [1], $|c_2| \leq \frac{2}{3}$ [4, 14], $|c_3| \leq \frac{1}{2} + e^{-6}$ [2]. It is an easy matter to deduce from Teichmüller's coefficient theorem that $|A_2| \leq 2$ and $|c_1| \leq 1$ but further efforts in this direction have been without success.

Of course these two bounds can be obtained by even the most elementary methods.

5. We will now give the first of our results which, although closely resembling Teichmüller's coefficient theorem, provide a great deal of information about the coefficients of univalent functions.

THEOREM 1. *Let $Q(w)dw^2$ be a quadratic differential on the w-sphere which is regular apart from a pole of order four at the point at infinity where Q has the expansion*

$$Q(w) = \alpha\left(1 + \frac{\beta_1}{w} + \text{higher powers of } w^{-1}\right)$$

and a simple pole at the origin. Let Δ be an admissible simply-connected domain with respect to $Q(w)dw^2$ which does not contain the origin and let f be a function univalent in Δ, not assuming the value zero and regular apart from a simple pole at the point at infinity where it has the expansion

$$f(w) = w + a_0 + \frac{a_1}{w} + \text{higher powers of } w^{-1}.$$

Then

$$\Re\{\alpha(a_1 + \beta_1 a_0)\} \leq 0. \tag{4}$$

Equality can occur in (4) *only if the mapping under f consists of a translation (in the Q-metric) along trajectories of $Q(w)dw^2$.*

This result is easily proved by the standard method of proof used for the General Coefficient Theorem. Further, numerous generalizations of it will be immediately evident. Nevertheless we have stated it in this special form and will give its proof in some detail in order to show how simply the present method leads to the coefficient results obtained even working from first principles.

We consider a sufficiently small simply-connected neighborhood U of the point at infinity and slit it along a trajectory Λ running in the one sense to the point at infinity and in the other leaving U. In the set $U - \overline{\Lambda}$ any particular determination of $\zeta = \int (Q(w))^{\frac{1}{2}} dw$ will be single-valued and map this set on a portion of Riemann surface over the ζ-plane. In the present situation the domains in the trajectory structure of $Q(w)dw^2$ are two end domains $\mathfrak{E}_1$, $\mathfrak{E}_2$ and a possible strip domain $\mathfrak{S}$ [10, Ch. III]. We regard in the ζ-plane a square of side $2L$, centre at the origin and with sides parallel to the real and imaginary axes. We take the trace of this square on the Riemann surface and its inverse image on the w-sphere. For L sufficiently large the latter is a sequence of arcs as follows. First there is an arc of an orthogonal trajectory with initial point on Λ. Then there is an arc of a trajectory lying entirely in one end domain. Next there is an arc of an orthogonal trajectory. Then there is an arc of a trajectory lying entirely in the other end domain. Finally there is an arc of an orthogonal trajectory ending on Λ. Its terminal point may not coincide with the initial point of the first arc above. Joining these points with an arc of Λ if necessary we obtain a simple closed curve which we denote by $\gamma(L)$. Removing the closure of the neighborhood of the point at infinity bounded by $\gamma(L)$ from Δ we obtain a domain which we denote by $\Delta(L)$. We will assume L so large that the boundary of Δ lies in the interior of the closure of $\Delta(L)$. Let the image of $\Delta(L)$ under f be denoted by $\Delta'(L)$. The proof of Theorem 1 is carried out by comparing the areas of $\Delta(L)$ and $\Delta'(L)$ in the Q-metric, i.e., the metric $|Q(w)|^{\frac{1}{2}}|dw|$.

We choose the determination of $\int(Q(w))^{\frac{1}{2}}dw$ above so that we have for the mapping of $U - \overline{\Lambda}$

$$\zeta = \int \alpha^{\frac{1}{2}}\left(1 + \frac{\beta_1}{w} + \frac{\beta_2}{w^2} + \ldots\right)^{\frac{1}{2}} dw$$
$$= \alpha^{\frac{1}{2}}w + \tfrac{1}{2}\alpha^{\frac{1}{2}}\beta_1 \log w - \alpha^{\frac{1}{2}}\frac{(\frac{1}{2}\beta_2 - \frac{1}{8}\beta_1^2)}{w} + \ldots .$$

Now the mapping f in the w-sphere induces a mapping in terms of the variable ζ. That is, starting with a value on the above Riemann surface we take the corresponding value w, perform the mapping $f(w)$ and map the image point by the same determination of $\int(Q(w))^{\frac{1}{2}}dw$ which we follow continuously. We denote by ω the value corresponding to ζ by this process. Then we have for ζ large enough

$$\omega = \alpha^{\frac{1}{2}}\left(w + a_0 + \frac{a_1}{w} + \ldots\right) + \tfrac{1}{2}\alpha^{\frac{1}{2}}\beta_1 \log\left(w + a_0 + \frac{a_1}{w} + \ldots\right)$$
$$- \alpha^{\frac{1}{2}}(\tfrac{1}{2}\beta_2 - \tfrac{1}{8}\beta_1^2)(w + a_0 + \ldots)^{-1} + \ldots$$
$$= \alpha^{\frac{1}{2}}w + \alpha^{\frac{1}{2}}a_0 + \frac{\alpha^{\frac{1}{2}}a_1}{w} + \ldots + \tfrac{1}{2}\alpha^{\frac{1}{2}}\beta_1 \log w + \frac{\frac{1}{2}\alpha^{\frac{1}{2}}\beta_1 a_0}{w} + \ldots$$
$$- \alpha^{\frac{1}{2}}\frac{(\frac{1}{2}\beta_2 - \frac{1}{8}\beta_1^2)}{w} + \ldots,$$

$$\omega = \zeta + \alpha^{\frac{1}{2}}a_0 + \frac{\alpha(a_1 + \frac{1}{2}\beta_1 a_0)}{\zeta} + O(|\zeta|^{-\frac{1}{2}}) . \tag{5}$$

We observe that the area of $\Delta'(L)$ in the Q-metric is bounded above by the area of the domain on the w-sphere bounded by the image of $\gamma(L)$ under f (not containing the point at infinity). We will estimate the difference of this quantity from the corresponding area of $\Delta(L)$. The change in area may be regarded as arising from the effect of the mapping on the boundary of a square adjusted by the effect on possible horizontal and vertical segments. The former contribution is seen at once to be

$$\Re\left\{\frac{1}{2i}\int(\bar{\zeta} + \bar{b}_0 + \bar{b}_1\bar{\zeta}^{-1})(1 - b_1\zeta^{-2})d\zeta - \frac{1}{2i}\int\bar{\zeta}d\zeta\right\} + O(L^{-\frac{1}{2}}),$$

where the integrals are taken over the boundary of the square and we have written b_0 for $\alpha^{\frac{1}{2}}a_0$, b_1 for $\alpha(a_1 + \frac{1}{2}\beta_1 a_0)$. The whole expression is seen at once to be $O(L^{-\frac{1}{2}})$. A circuit in the z-plane about the point at infinity returns us in the Riemann surface over the ζ-plane to a point whose affix has been translated by $\pi i\alpha^{\frac{1}{2}}\beta_1$. Since up to terms which are $O(L^{-1})$ the transformation from ζ to ω is a translation through $\alpha^{\frac{1}{2}}a_0$ the contribution from the segments is seen to be

$$\Im\{\pi i\alpha^{\frac{1}{2}}\beta_1\bar{\alpha}^{\frac{1}{2}}\bar{a}_0\} + O(L^{-1}) = \pi\Re\{|\alpha|\beta_1\bar{a}_0\} + O(L^{-1}).$$

Thus the total change in area is

$$\pi\Re\{|\alpha|\beta_1\bar{a}_0\} + O(L^{-\frac{1}{2}})$$

and we have

$$\iint_{\Delta'(L)} dA \leq \iint_{\Delta(L)} dA + \pi\Re\{|\alpha|\beta_1\bar{a}_0\} + O(L^{-\frac{1}{2}}) \tag{6}$$

where dA denotes the element of area in the Q-metric.

To estimate the area of $\Delta'(L)$ from below we interpret it as the area of $\Delta(L)$ in a new metric

$$\rho(w)|Q(w)|^{\frac{1}{2}}|dw| = |f'_{\cdot}(w)||Q(w)|^{\frac{1}{2}}|dw|$$

and apply the method of the extremal metric. We denote the intersections of $\mathfrak{E}_1$, $\mathfrak{E}_2$, and $\mathfrak{S}$ (if present) with $\Delta(L)$ by $\mathfrak{E}_1(L)$, $\mathfrak{E}_2(L)$, and $\mathfrak{S}(L)$. The image of one of the former, say $\mathfrak{E}_1(L)$, under the given branch of $\int(Q(w))^{\frac{1}{2}}dw$ is the rectangle $E(L)$ defined by ($\zeta = \xi + i\eta$)

$$-L < \xi < L, \quad -\lambda < \eta < L$$

where λ is independent of L and may be positive or negative. We transfer the given metric to the ζ-plane by setting

$$\rho(\zeta)|d\zeta| = \rho(w)|Q(w)|^{\frac{1}{2}}|dw|.$$

No confusion arises by using the notation ρ in each case.

Let now $\sigma(\eta)$ denote the horizontal segment of length $2L$ lying in $E(L)$ on the line $\Im\zeta = \eta(-\lambda < \eta < L)$ and let $s(\eta)$ be its antecedent in the w-sphere. Since $\int_{\sigma(\eta)} \rho d\xi$ is equal to the length in the Q-metric of the image of $s(\eta)$ under f we easily see from the expansion (5) that

$$\int_{\sigma(\eta)} \rho d\xi \geq 2L + 2ML^{-1}\cos\mu\cos^2\theta + O(L^{-\frac{3}{2}})$$

where we have set

$$\eta = L\tan\theta \quad \left(-\frac{\pi}{4} < \theta < \frac{\pi}{4}\right)$$

$$\alpha(a_1 + \tfrac{1}{2}\beta_1 a_0) = Me^{i\mu} \quad (M \geq 0, \mu \text{ real}).$$

Integrating this with respect to η over the range $-\lambda < \eta < L$ we get

$$\iint\limits_{E(L)} \rho d\xi d\eta \geq 2L(L+\lambda) + 2ML^{-1}\cos\mu \int_{\theta_0}^{\pi/4} \cos^2\theta dL \tan\theta + O(L^{-\frac{1}{2}})$$

where θ_0 is the value of the angle θ corresponding to $\eta = -\lambda$. Replacing it by zero changes the second term on the right hand side by only an amount $O(L^{-1})$. It is evident that

$$\iint\limits_{\mathfrak{E}_1(L)} \rho dA = \iint\limits_{E(L)} \rho d\xi d\eta.$$

Thus

$$\iint\limits_{\mathfrak{E}_1(L)} \rho dA \geq \iint\limits_{\mathfrak{E}_1(L)} dA + \frac{\pi}{2}\,\Re\{\alpha(a_1 + \tfrac{1}{2}\beta_1 a_0)\} + O(L^{-\frac{1}{2}}). \tag{7}$$

On the other hand

$$0 \leq \iint\limits_{\mathfrak{E}_1(L)} (\rho - 1)^2 dA = \iint\limits_{\mathfrak{E}_1(L)} \rho^2 dA + \iint\limits_{\mathfrak{E}_1(L)} dA - 2\iint\limits_{\mathfrak{E}_1(L)} \rho dA$$

so that

$$\iint\limits_{\mathfrak{E}_1(L)} \rho^2 dA \geq 2\iint\limits_{\mathfrak{E}_1(L)} \rho dA - \iint\limits_{\mathfrak{E}_1(L)} dA$$

$$\geq \iint\limits_{\mathfrak{E}_1(L)} dA + \pi\Re\{\alpha(a_1 + \tfrac{1}{2}\beta_1 a_0)\} + O(L^{-\frac{1}{2}}).$$

It follows similarly that

$$\iint\limits_{\mathfrak{E}_2(L)} \rho^2 dA \geq \iint\limits_{\mathfrak{E}_2(L)} dA + \pi\Re\{\alpha(a_1 + \tfrac{1}{2}\beta_1 a_0)\} + O(L^{-\frac{1}{2}}).$$

If a strip domain $\mathfrak{S}$ is present, it is verified that

$$\iint\limits_{\mathfrak{S}(L)} \rho^2 dA \geq \iint\limits_{\mathfrak{S}(L)} dA + 2\Re\{\pi\alpha^{\frac{1}{2}}\beta_1\}\Re\{\alpha^{\frac{1}{2}}a_0\} + O(L^{-1}).$$

Combining these inequalities we have

$$\iint\limits_{\Delta'(L)} dA = \iint\limits_{\Delta(L)} \rho^2 dA \geq \iint\limits_{\Delta(L)} dA + 2\pi\Re\{\alpha(a_1 + \tfrac{1}{2}\beta_1 a_0)\}$$
$$+ 2\Re\{\pi\alpha^{\frac{1}{2}}\beta_1\}\Re\{\alpha^{\frac{1}{2}}a_0\} + O(L^{-\frac{1}{2}}).$$

Comparing this with inequality (6) we obtain at once inequality (4).

To justify the equality statement we observe that unless the mapping by f carries every trajectory of $Q(w)dw^2$ into another such it follows by a standard method [7; 10, Ch. II] that a positive constant term can be inserted on the right hand side of inequality (7). This implies at once that the inequality (4) is strict. On the other hand, since in the present situation the origin is a boundary point of Δ, for equality to occur in (4) the mapping under f could only be a translation (in the Q-metric) along trajectories of $Q(w)dw^2$.

6. LEMMA 1. *Let $Q_\phi(w)dw^2$ be the quadratic differential*

$$e^{-2i\phi}(\tau e^{i\phi} - w)w^{-1}dw^2 \quad (\phi \textit{ real}, 0 < \tau \leq 4).$$

Let $f(z, \tau, \phi)$ be the function in Σ mapping $|z| > 1$ conformally onto the domain Δ_ϕ bounded by the rectilinear segment from the origin to $\tau e^{i\phi}$ together with possible arcs of equal length on the closures of the other trajectories of $Q_\phi(w)dw^2$ having limiting end points at $\tau e^{i\phi}$ (these arcs being absent when $\tau = 4$). Then the expansion of $f(z, \tau, \phi)$ about the point at infinity begins

$$z + \tfrac{1}{2}\tau[1 - \log(\tau/4)]e^{i\phi} + \frac{[\frac{1}{8}\tau^2(1 - 2\log(\tau/4)) - 1]e^{2i\phi}}{z} + \cdots.$$

When $0 < \tau < 4$ there can be obtained from $f(z, \tau, \phi)$ by translation along the trajectories of $Q_\phi(w)dw^2$ a one parameter family of functions $F(z, \tau, \phi, \lambda)$ where λ denotes the (sensed) amount of translation (in the Q_ϕ-metric) and satisfies the inequalities

$$-2(1 - \tfrac{1}{16}\tau^2)^{\frac{1}{2}} + \tfrac{1}{2}\tau \cos^{-1}\frac{\tau}{4} \leq \lambda \leq 2(1 - \tfrac{1}{16}\tau^2)^{\frac{1}{2}} - \tfrac{1}{2}\tau \cos^{-1}\frac{\tau}{4}.$$

In particular

$$F(z, \tau, \phi, 0) \equiv f(z, \tau, \phi).$$

The expansion of $F(z, \tau, \phi, \lambda)$ about the point at infinity begins

$$z + [i\lambda + \tfrac{1}{2}\tau(1 - \log(\tau/4))]e^{i\phi} + \frac{[\frac{1}{2}i\lambda\tau + \frac{1}{8}\tau^2 - \frac{1}{4}\tau^2\log(\tau/4) - 1]e^{2i\phi}}{z} + \cdots.$$

For $\tau = 4$ there are no functions obtained from $f(z, 4, \phi)$ by translation along the trajectories of $Q_\phi(w)dw^2$.

Consider first the case $\phi = 0$. Then for a suitable choice of determination the mapping

$$\zeta = \int_\tau^w \left(\frac{w - \tau}{w}\right)^{\frac{1}{2}} dw$$

carries the upper half w-plane onto a domain bounded by the half-infinite segment

$$\Im\zeta = -\tfrac{1}{2}\pi\tau, \quad -\infty < \Re\zeta \leq 0,$$

the segment

$$\Re\zeta = 0, \quad -\tfrac{1}{2}\pi\tau \leq \Im\zeta \leq 0$$

and the positive real axis, the image domain lying above this boundary in the ζ-plane. This mapping is given explicitly by the formula

$$\zeta = [w(w-\tau)]^{\frac{1}{2}} + \tfrac{1}{2}\log\left[\frac{w^{\frac{1}{2}} - (w-\tau)^{\frac{1}{2}}}{w^{\frac{1}{2}} + (w-\tau)^{\frac{1}{2}}}\right], \tag{8}$$

where the roots have their positive determination for w large and positive and we take the principal branch of the logarithm there.

Next the mapping

$$\zeta = \int_1^z z^{-2}(z - e^{i\theta})(z - e^{-i\theta})\,dz$$

with $0 \le \theta < \dfrac{\pi}{2}$ carries the domain $\Im z > 0$, $|z| > 1$ onto a domain bounded by the half-infinite segment

$$\Im\zeta = -2\pi\cos\theta, \quad -\infty < \Re\zeta \le 0,$$

the segment

$$\Re\zeta = 0, \quad -2\pi\cos\theta \le \Im\zeta \le 2\sin\theta - 2\theta\cos\theta$$

and the positive real axis, the image domain lying above this boundary in the ζ-plane. This mapping is given explicitly by the formula

$$\zeta = z - 2\cos\theta\log z - z^{-1}, \tag{9}$$

where we take the principal branch of the logarithm for z large and positive.

If we choose

$$\cos\theta = \tfrac{1}{4}\tau$$

the mappings (8) and (9) combine to give a conformal mapping from the domain $\Im z > 0$, $|z| > 1$ onto the upper half w-plane slit (except for $\tau = 4$) along an arc on the trajectory of $Q_0(w)dw^2$ contained therein which has a limiting end point at τ. By reflection this mapping extends to a conformal mapping $f(z, \tau, 0)$ of $|z| > 1$ which is seen at once to be in Σ and to carry $|z| > 1$ onto the domain Δ_0. Inserting a trial expansion

$$w = z + a_0 + \frac{a_1}{z} + \ldots$$

in (8) we obtain

$$\zeta = z - \tfrac{1}{2}\tau\log z + [a_0 - \tfrac{1}{2}\tau(1 - \log(\tau/4))] + \frac{a_1 + \frac{1}{8}\tau^2 - \frac{1}{2}\tau a_0}{z} + \ldots. \tag{10}$$

Comparing this with (9) we have

$$a_0 = \tfrac{1}{2}\tau(1 - \log(\tau/4))$$

$$a_1 = \tfrac{1}{8}\tau^2(1 - 2\log(\tau/4)) - 1.$$

Obtaining a function from $f(z, \tau, 0)$ by a translation of amount λ (in the Q_0-metric) along the trajectories of $Q_0(w)dw^2$ corresponds to translating the ζ value corresponding to w by $i\lambda$ and repeating the preceding process. This will still lead to a function in Σ provided that

$$-2(1 - \tfrac{1}{16}\tau^2)^{\frac{1}{2}} + \tfrac{1}{2}\tau \cos^{-1}\frac{\tau}{4} \leq \lambda \leq 2(1 - \tfrac{1}{16}\tau^2)^{\frac{1}{2}} - \tfrac{1}{2}\tau \cos^{-1}\frac{\tau}{4}.$$

For $\tau = 4$ there are no such functions (except for $\lambda = 0$). For $0 < \tau < 4$ there is a one parameter family of such functions which we denote by $F(z, \tau, 0, \lambda)$. The expansions of these functions about the point at infinity are obtained by equating (10) not with (9) but rather with

$$\zeta = z + i\lambda - 2\cos\theta \log z - z^{-1}.$$

This gives for the coefficients

$$a_0 = i\lambda + \tfrac{1}{2}\tau(1 - \log(\tau/4))$$

$$a_1 = \tfrac{1}{2}i\lambda\tau + \tfrac{1}{8}\tau^2 - \tfrac{1}{4}\tau^2 \log(\tau/4) - 1.$$

Now to deal with the case of general real ϕ we form the respective functions

$$f(z, \tau, \phi) = e^{i\phi}f(ze^{-i\phi}, \tau, 0)$$

$$F(z, \tau, \phi, \lambda) = e^{i\phi}F(ze^{-i\phi}, \tau, 0, \lambda)$$

the latter applying only in the case $0 < \tau < 4$. This completes the proof of Lemma 1.

LEMMA 2. *If the function $f(z) \in S$ has the expansion* (2) *about the origin then the function $1/f(1/z)$ is in Σ, does not take the value zero for $|z| > 1$ and has expansion about the point at infinity.*

$$z - A_2 + \frac{A_2^2 - A_3}{z} + \frac{-A_2^3 + 2A_2A_3 - A_4}{z^2} + \cdots.$$

This is obvious.

7. COROLLARY 1. *Let $Q(w)dw^2$ be a quadratic differential on the w-sphere which is regular apart from a pole of order four at the point at infinity where Q has the expansion*

$$Q(w) = \alpha\left(1 + \frac{\beta_1}{w} + \text{higher powers of } w^{-1}\right)$$

and a simple pole at the origin. Let $f^*(z) \in \Sigma$ and map $|z| > 1$ onto a domain Δ admissible with respect to $Q(w)dw^2$ and not containing the origin. Let $f^*(z)$ have the expansion about the point at infinity

$$f^*(z) = z + b_0 + \frac{b_1}{z} + \cdots.$$

Let $f(z) \in \Sigma$, not take the value zero for $|z| > 1$ and have the expansion (3) *about the point at infinity. Then*

$$\Re\{\alpha(c_1 + \beta_1 c_0)\} \leq \Re\{\alpha(b_1 + \beta_1 b_0)\}. \tag{11}$$

Equality can occur in (11) *only if $f(z)$ is obtained from $f^*(z)$ by translation along the trajectories of $Q(w)dw^2$.*

Let $\psi(w)$ be the inverse function of $f^*(z)$ defined in Δ. Then Theorem 1 can be applied to the function.

$$g(w) = f(\psi(w)).$$

The expansion of this function about the point at infinity is

$$g(w) = w + (c_0 - b_0) + \frac{(c_1 - b_1)}{w} + \dots.$$

From inequality (4) thus follows

$$\Re\{\alpha[(c_1 - b_1) + \beta_1(c_0 - b_0)]\} \leq 0$$

which is equivalent to inequality (11). The equality statement follows at once from Theorem 1.

COROLLARY 2. *Let $f(z) \in \Sigma$, not take the value zero for $|z| > 1$ and have the expansion* (3) *about the point at infinity. Then for ϕ real and $0 \leq \tau \leq 4$*

$$\begin{aligned} \Re\{e^{-2i\phi}(c_1 - \tau e^{i\phi} c_0)\} &\geq -1 - \tfrac{3}{8}\tau^2 + \tfrac{1}{4}\tau^2 \log(\tau/4), \quad 0 < \tau \leq 4 \\ &\geq -1 \quad\quad\quad\quad\quad\quad\quad\quad\quad\quad\quad\;, \quad \tau = 0. \end{aligned} \tag{12}$$

For $\tau = 0$ equality occurs in (12) *only for the functions $z + iae^{i\phi} - e^{2i\phi}z^{-1}$ where $-2 \leq a \leq 2$. For $0 < \tau < 4$ equality occurs in* (12) *only for the functions $F(z, \tau, \phi, \lambda)$. For $\tau = 4$ equality occurs in* (12) *only for the function $f(z, \tau, \phi) \equiv z + 2e^{i\phi} + e^{2i\phi}z^{-1}$.*

The results in the case $\tau = 0$ follow at once from Teichmüller's coefficient theorem, keeping in mind the fact that $f(z)$ is not to take the value zero for $|z| > 1$ in dealing with the case of equality.

For $0 < \tau \leq 4$ we apply Corollary 1 with

$$Q(w)dw^2 = e^{-2i\phi}(\tau e^{i\phi} - w)w^{-1}dw^2$$
$$f^*(z) = f(z, \tau, \phi)$$

so that

$$\begin{aligned} \alpha &= -e^{-2i\phi} \\ \beta_1 &= -\tau e^{i\phi} \\ b_0 &= \tfrac{1}{2}\tau[1 - \log(\tau/4)]e^{i\phi} \\ b_1 &= [\tfrac{1}{8}\tau^2(1 - 2\log(\tau/4)) - 1]e^{2i\phi}. \end{aligned}$$

Inserting these values in inequality (11) we obtain inequality (12). The equality statements follow from that of Corollary 1 and the enumeration in Lemma 1 of the functions which can be obtained from $f(z, \tau, \phi)$ by translation along the trajectories of $Q_\phi(w)dw^2$.

COROLLARY 3. *Let $f(z) \in S$ and have the expansion* (2) *about the origin. Then for ϕ real and $0 \leq \tau \leq 4$*

$$(13)\quad \Re\{-e^{-2i\phi}A_3 + e^{-2i\phi}A_2^2 + \tau e^{-i\phi}A_2\} \geq -1 - \tfrac{3}{8}\tau^2 + \tfrac{1}{4}\tau^2 \log(\tau/4), \quad 0 < \tau \leq 4$$

$$\geq -1 \qquad , \quad \tau = 0.$$

For $\tau = 0$ equality occurs in (13) *only for the functions $z(1 + iae^{i\phi}z - e^{2i\phi}z^2)$ where $-2 \leq a \leq 2$.* For $0 < \tau < 4$ *equality occurs in* (13) *only for the functions $1/F(1/z, \tau, \phi, \lambda)$. For $\tau = 4$ equality occurs in* (13) *only for the function $1/f(1/z, 4, \phi) \equiv z(1 + e^{i\phi}z)^{-2}$.*

This follows at once from Lemma 2 and Corollary 2.

COROLLARY 4. *Let $f(z) \in S$ and have the expansion* (2) *about the origin. Let ϕ be real. If*

$$\Re\{e^{-i\phi}A_2\} = 0$$

then

$$(14)\qquad \Re\{e^{-2i\phi}A_3\} \leq 1.$$

Equality occurs in (14) *only for the function $z(1 - e^{2i\phi}z^2)^{-1}$. If*

$$\Re\{e^{-i\phi}A_2\} = -\tfrac{1}{2}\tau(1 - \log(\tau/4)), \quad 0 < \tau \leq 4,$$

then

$$(15)\qquad \Re\{e^{-2i\phi}A_3\} \leq 1 + \tfrac{1}{8}\tau^2 - \tfrac{1}{4}\tau^2 \log(\tau/4) + \tfrac{1}{4}\tau^2[\log(\tau/4)]^2.$$

Equality occurs in (15) *only for the function $1/f(1/z, \tau, \phi)$.*

From inequality (13) follows that

$$(16)\quad \Re\{e^{-2i\phi}A_3\} \leq 1 + \tfrac{3}{8}\tau^2 - \tfrac{1}{4}\tau^2 \log(\tau/4) + \tau\Re\{e^{-i\phi}A_2\} + \Re\{e^{-2i\phi}A_2^2\}, \quad 0 < \tau \leq 4$$

$$\leq 1 + \Re\{e^{-2i\phi}A_2^2\} \qquad , \quad \tau = 0.$$

Now

$$(17)\qquad \Re\{e^{-2i\phi}A_2^2\} \leq [\Re\{e^{-i\phi}A_2\}]^2$$

the inequality being strict unless

$$\Im\{e^{-i\phi}A_2\} = 0.$$

If

$$\Re\{e^{-i\phi}A_2\} = 0$$

taking $\tau = 0$ in (16) we have

$$\Re\{e^{-2i\phi}A_3\} \leq 1$$

with equality occurring only for the function $z(1 - e^{2i\phi}z^2)^{-1}$ in view of the preceding remark. If

$$\Re\{e^{-i\phi}A_2\} = -\tfrac{1}{2}\tau(1 - \log(\tau/4)), \quad 0 < \tau \leq 4$$

from (16) and (17) follows

$$\Re\{e^{-2i\phi}A_3\} \leq 1 + \tfrac{1}{8}\tau^2 - \tfrac{1}{4}\tau^2(1 - \log(\tau/4)) + \tfrac{1}{4}\tau^2(1 - \log(\tau/4))^2$$

which reduces at once to (15). Equality can occur there only for the function $1/f(1/z, \tau, \phi)$ in view of the equality conditions for (13) and (17).

COROLLARY 5. *Let $f(z) \in S$ and have the expansion* (2) *about the origin. If*

$$|A_2| = 0$$

then

$$|A_3| \leq 1. \tag{18}$$

Equality occurs in (18) *only for the functions $z(1 - e^{2i\phi}z^2)^{-1}$, ϕ real. If*

$$|A_2| \leq \tfrac{1}{2}\tau(1 - \log(\tau/4)), \quad 0 < \tau \leq 4,$$

then

$$|A_3| \leq 1 + \tfrac{1}{8}\tau^2 - \tfrac{1}{4}\tau^2 \log(\tau/4) + \tfrac{1}{4}\tau^2[\log(\tau/4)]^2. \tag{19}$$

Equality occurs in (19) *only for the functions $1/f(1/z, \tau, \phi)$, ϕ real.*

If

$$|A_2| = 0$$

we can choose ϕ so that

$$\Re\{e^{-2i\phi}A_3\} = |A_3|.$$

Then inequality (18) follows from (14). The equality statement is immediate.

In any case we can choose ϕ so that

$$\Re\{e^{-2i\phi}A_3\} = |A_3|$$

while

$$\Re\{e^{-i\phi}A_2\} \leq 0.$$

If

$$\Re\{e^{-i\phi}A_2\} = 0$$

the bound (18) applies. Otherwise we can find τ, $0 < \tau \leq 4$, so that

$$\Re\{e^{-i\phi}A_2\} = -\tfrac{1}{2}\tau[1 - \log(\tau/4)].$$

Using this value of τ we obtain from (15)

$$|A_3| \leq 1 + \tfrac{1}{8}\tau^2 - \tfrac{1}{4}\tau^2 \log(\tau/4) + \tfrac{1}{4}\tau^2[\log(\tau/4)]^2.$$

The function on the right hand side of this inequality has derivative $\tfrac{1}{2}\tau[\log(\tau/4)]^2$ with respect to τ and thus is strictly increasing as a function of τ for $0 < \tau \leq 4$. As τ tends to zero this function has limit one. We note also that the function $\frac{1}{2}\tau[1 - \log(\tau/4)]$ is strictly increasing as a function of τ for $0 < \tau \leq 4$. Thus if

$$|A_2| \leq \tfrac{1}{2}\tau[1 - \log(\tau/4)]$$

the bound (19) applies. Equality can occur here only if it does in (15), thus only for the functions $1/f(1/z, \tau, \phi)$, ϕ real.

COROLLARY 6. *Let $f(z) \in S$ and have the expansion* (2) *about the origin. Then*

$$|A_3| \leq 3. \tag{20}$$

Equality occurs in (20) *only for the functions $z(1 + e^{i\phi}z)^{-2}$, ϕ real.*

This result is immediate, taking $\tau = 4$ in Corollary 5.

COROLLARY 7. *Let $f(z) \in S$ and have the expansion* (2) *about the origin. Then*

$$|A_3| - |A_2| \leq \tfrac{1}{8}\tau_0^2 - \tfrac{1}{4}\tau_0 + \tfrac{3}{4} \tag{21}$$

where τ_0 is the larger root of the equation

$$\tau \log(\tau/4) + 1 = 0$$

on the interval $0 < \tau < 4$. Equality occurs in (21) *only for the functions $1/f(1/z, \tau_0, \phi)$, ϕ real.*

This result is due in a slightly different form to Golusin [5]. His lower bound

$$-1 \leq |A_3| - |A_2|$$

is elementary.

We define

$$\begin{aligned}\Psi(\tau) &= 1 + \tfrac{1}{8}\tau^2 - \tfrac{1}{4}\tau^2\log(\tau/4) + \tfrac{1}{4}\tau^2[\log(\tau/4)]^2 - \tfrac{1}{2}\tau[1 - \log(\tau/4)], \quad 0 < \tau \leq 4\\ &= 1 \qquad\qquad , \quad \tau = 0.\end{aligned}$$

Then the function $\Psi(\tau)$ is continuous on $0 \leq \tau \leq 4$. From Corollary 5 follows

$$|A_3| - |A_2| \leq \max_{0 \leq \tau \leq 4} \Psi(\tau).$$

Now for $0 < \tau \leq 4$

$$\frac{d\Psi(\tau)}{d\tau} = \tfrac{1}{2}[\tau \log(\tau/4) + 1]\log(\tau/4).$$

This last function is found to have two zeros τ_0', τ_0, $\tau_0' < \tau_0$ on $0 < \tau < 4$. The function Ψ has the value one at $\tau = 0$, decreases for $0 < \tau < \tau_0'$ increases for $\tau_0' < \tau < \tau_0$, decreases for $\tau_0 < \tau < 4$ and has the value one at $\tau = 4$. Thus its maximum on [0, 4] occurs at τ_0. We verify at once that

$$\Psi(\tau_0) = \tfrac{1}{8}\tau_0^2 - \tfrac{1}{4}\tau_0 + \tfrac{3}{4}.$$

The equality statement of Corollary 7 follows from the fact that equality can occur in (21) only if it occurs in (19) for the value $\tau = \tau_0$.

COROLLARY 8. *Let $f(z) \in S$ and have the expansion* (2) *about the origin. Then for $\mu > 1$*

$$|A_3 - \mu A_2^2| \leq 4\mu - 3. \tag{22}$$

Equality occurs in (22) *only for the functions $z(1 + e^{i\phi}z)^{-2}$, ϕ real. For $\mu = 1$*

$$|A_3 - A_2^2| \leq 1. \tag{23}$$

Equality occurs in (23) *only for the functions $z(1 + ce^{i\phi} + e^{2i\phi}z^2)^{-1}$, ϕ real, $-2 \leq c \leq 2$. For $0 < \mu < 1$*

$$|A_3 - \mu A_2^2| \leq 1 + 2 \exp\left(-\frac{2\mu}{1-\mu}\right). \tag{24}$$

Equality occurs in (24) *only for the functions $1/f\left(1/z, 4\exp\left(-\frac{\mu}{1-\mu}\right), \phi\right)$, ϕ real. For $\mu \leq 0$*

$$|A_3 - \mu A_2^2| \leq 3 - 4\mu. \tag{25}$$

Equality occurs in (25) *only for the functions $z(1 + e^{i\phi}z)^{-2}$, ϕ real.*

This result is in the main due to Golusin [6], some special cases having been given earlier by other authors.

Inequality (23) is a well known elementary result which can be derived immediately from Corollary 3 by taking $\tau = 0$, together with the corresponding equality statement. Then for $\mu > 1$

$$\begin{aligned} |A_3 - \mu A_2^2| &\leq |A_3 - A_2^2| + (\mu - 1)|A_2|^2 \\ &\leq 1 + 4(\mu - 1) = 4\mu - 3. \end{aligned}$$

Equality can occur here only for functions for which

$$|A_2| = 2,$$

i.e. the functions $z(1 + e^{i\phi}z)^{-2}$, ϕ real.

Now let μ be less than one. By Corollary 3 we have for any real ϕ

$$\begin{aligned} \Re\{e^{-2i\phi}(A_3 - \mu A_2^2)\} &\leq 1 + \tfrac{3}{8}\tau + (1-\mu)\Re\{e^{-2i\phi}A_2^2\} \\ &\quad + \tau\Re\{e^{-i\phi}A_2\}, \quad 0 \leq \tau \leq 4 \\ &\leq 1 + (1-\mu)\Re\{e^{-2i\phi}A_2^2\}, \quad \tau = 0. \end{aligned} \tag{26}$$

We can choose ϕ so that

$$\Re\{e^{-2i\phi}(A_3 - \mu A_2^2)\} = |A_3 - \mu A_2^2|$$

while

$$\Re\{e^{-i\phi}A_2\} \leq 0.$$

If

$$\Re\{e^{-i\phi}A_2\} = 0$$

taking $\tau = 0$ in (26) and using inequality (17) we have

$$|A_3 - \mu A_2^2| \leq 1.$$

Otherwise we can find τ, $0 < \tau \leq 4$, so that

$$\Re\{e^{-i\phi}A_2\} = -\tfrac{1}{2}\tau[1 - \log(\tau/4)].$$

Using this value of τ and inequality (17) in (26) we obtain

$$|A_3 - \mu A_2^2| \leq 1 + \frac{(1-2\mu)}{8}\tau^2 + \frac{(2\mu-1)}{4}\tau^2 \log(\tau/4) + \frac{(1-\mu)}{4}\tau^2[\log(\tau/4)]^2.$$

Now we define

$$\Phi_\mu(\tau) = 1 + \frac{(1-2\mu)}{8}\tau^2 + \frac{(2\mu-1)}{4}\tau^2 \log(\tau/4) + \frac{(1-\mu)}{4}\tau^2[\log(\tau/4)]^2, \quad 0 < \tau \leq 4$$

$$= 1 \qquad , \quad \tau = 0.$$

The function $\Phi_\mu(\tau)$ is continuous on $0 \leq \tau \leq 4$. Further we have

$$|A_3 - \mu A_2^2| \leq \max_{0 \leq \tau \leq 4} \Phi_\mu(\tau).$$

Now for $0 < \tau \leq 4$

$$\frac{d\Phi_\mu(\tau)}{d\tau} = \tau \log(\tau/4)\left[\frac{1-\mu}{2}\log(\tau/4) + \frac{\mu}{2}\right].$$

Thus for $\mu \leq 0$ the function $\Phi_\mu(\tau)$ is an increasing function on [0, 4], its maximum occurs for $\tau = 4$ and

$$|A_3 - \mu A_2^2| \leq 1 + 2(1 - 2\mu) = 3 - 4\mu.$$

Since we have used Corollary 3 equality can occur here only for the functions $1/f(1/z, 4, \phi) \equiv z(1 + e^{i\phi}z)^{-2}$, ϕ real. For $0 < \mu < 1$, the function $\Phi_\mu(\tau)$ increases from $\tau = 0$ up to the place where

$$\frac{1-\mu}{2}\log(\tau/4) + \frac{\mu}{2} = 0$$

that is

$$\tau = 4\exp\left(-\frac{\mu}{1-\mu}\right)$$

and then decreases. Thus the maximum of $\Phi_\mu(\tau)$ occurs for this value of τ and

$$|A_3 - \mu A_2^2| \leq 1 + 2\exp\left(-\frac{2\mu}{1-\mu}\right).$$

Since we have used Corollary 3 and inequality (17) equality can occur here only for the functions $1/f\left(1/z, 4\exp\left(-\frac{\mu}{1-\mu}\right), \phi\right)$, ϕ real.

Although the preceding results are contained implicitly in the characterization of the coefficient region for A_2, A_3 given by Schaeffer and Spencer [13] in most cases it does not seem very easy to derive them explicitly from that source.

8. THEOREM 2. *Let $Q(w)dw^2$ be a quadratic differential on the w-sphere which is regular apart from a pole of order five at the point at infinity where Q has the expansion*

$$Q(w) = \alpha(w + \beta_0 + \text{terms in } w^{-1}).$$

Let Δ be an admissible simply-connected domain with respect to $Q(w)dw^2$ and let f be a function univalent in Δ and regular apart from a simple pole at the point at infinity where it has the expansion

$$f(w) = w + a_0 + \frac{a_1}{w} + \frac{a_2}{w^2} + \dots .$$

Then

$$\Re\{\alpha(a_2 + \beta_0 a_1)\} \leq 0. \tag{27}$$

Equality can occur in (27) *only if the mapping under f is a* (*Euclidean*) *translation.*

Once again this result is readily seen to admit numerous generalizations. The proof follows on the same lines as that of Theorem 1 but some new considerations intervene.

We consider a sufficiently small simply-connected neighborhood U of the point at infinity and slit it along a trajectory Λ of $Q(w)dw^2$ running in the one sense to the point at infinity and in the other leaving U. In the set $U - \overline{\Lambda}$ any particular determination of $\zeta = \int (Q(w))^{\frac{1}{2}}dw$ will be single-valued and map this set on a portion of Riemann surface over the ζ-plane. In the present case it is readily seen that the domains in the trajectory structure of $Q(w)dw^2$ are three end domains $\mathfrak{E}_1$, $\mathfrak{E}_2$, $\mathfrak{E}_3$ which we suppose occur in cyclic order about the point at infinity. We may suppose that the above determination of $\int (Q(w))^{\frac{1}{2}}dw$ maps them respectively onto an upper half-plane E_1, a lower half-plane E_2 and an upper half-plane E_3. Further we may take Λ as the common boundary arc of $\mathfrak{E}_1$ and $\mathfrak{E}_3$ and suppose that its image on the boundary of E_1 is the positive real axis, its image on the boundary of E_3 is the negative real axis. In the present instance it is technically advantageous to regard in the ζ-plane, instead of a square as in the proof of Theorem 1, a circle centre the origin and of radius R. The trace of this circle on the Riemann surface consists of three semicircles and its inverse image on the w-sphere runs from a point of Λ back to the same point (in this special case). We denote the simple closed curve so obtained by $\gamma(R)$. Removing the closure of the neighborhood of the point at infinity bounded by $\gamma(R)$ from Δ we obtain a domain which we denote by $\Delta(R)$. We will assume R so large that the boundary of Δ lies in the interior of the closure of $\Delta(R)$. Let the image of $\Delta(R)$ under f be denoted by $\Delta'(R)$. The proof is again carried out by comparing the areas of $\Delta(R)$ and $\Delta'(R)$ in the Q-metric.

With the given choice of determination of $\int (Q(w))^{\frac{1}{2}} dw$ we have for the mapping of $U - \overline{\Lambda}$

$$\zeta = \int \alpha^{\frac{1}{2}} \left[w + \beta_0 + \frac{\beta_1}{w} + \ldots \right]^{\frac{1}{2}} dw$$

$$= \tfrac{2}{3}\alpha^{\frac{1}{2}} w^{\frac{3}{2}} + \alpha^{\frac{1}{2}}\beta_0 w^{\frac{1}{2}} + k_0 - \alpha^{\frac{1}{2}}(\beta_1 - \tfrac{1}{4}\beta_0^2) w^{-\frac{1}{2}} + \text{higher powers of } w^{-\frac{1}{2}}$$

where k_0 is a suitable constant and $\alpha^{\frac{1}{2}}$ a proper choice of the root. Once again the mapping f in the w-sphere induces a mapping in terms of the variable ζ. We denote by ω the value corresponding to ζ by this process. Then we have for ζ large enough (where we assume as we may that the coefficient a_0 in the expansion of $f(w)$ is zero)

$$\omega = \tfrac{2}{3}\alpha^{\frac{1}{2}} \left(w + \frac{a_1}{w} + \frac{a_2}{w^2} + \ldots \right)^{\frac{3}{2}} + \alpha^{\frac{1}{2}}\beta_0 \left(w + \frac{a_1}{w} + \frac{a_2}{w^2} + \ldots \right)^{\frac{1}{2}} + k_0$$

$$- \alpha^{\frac{1}{2}}(\beta_1 - \tfrac{1}{4}\beta_0^2) \left(w + \frac{a_1}{w} + \frac{a_2}{w^2} + \ldots \right)^{-\frac{1}{2}} + \ldots$$

$$= \zeta + \alpha^{\frac{1}{2}} a_1 w^{-\frac{1}{2}} + \alpha^{\frac{1}{2}}(a_2 + \tfrac{1}{2}\beta_0 a_1) w^{-\frac{3}{2}} + \ldots .$$

On the other hand

$$w^{-\frac{3}{2}} = \tfrac{2}{3}\alpha^{\frac{1}{2}}\zeta^{-1} \left(1 + \frac{\frac{3}{2}\beta_0}{w} + \frac{\frac{3}{2}k_0}{w^{\frac{3}{2}}}\alpha^{-\frac{1}{2}} + \ldots \right)$$

and

$$w^{-\frac{1}{2}} = (\tfrac{2}{3})^{\frac{1}{3}}\alpha^{\frac{1}{6}}\zeta^{-\frac{1}{3}} \left(1 + \frac{\frac{3}{2}\beta_0}{w} + \ldots \right)^{\frac{1}{3}}$$

$$= (\tfrac{2}{3})^{\frac{1}{3}}\alpha^{\frac{1}{6}}\zeta^{-\frac{1}{3}} + \tfrac{1}{3}\alpha^{\frac{1}{2}}\beta_0\zeta^{-1} + \ldots$$

with appropriate choices of the roots. Thus finally

$$\omega = \zeta + \frac{(\frac{2}{3})^{\frac{1}{3}}\alpha^{\frac{2}{3}} a_1}{\zeta^{\frac{1}{3}}} + \frac{\frac{2}{3}\alpha(a_2 + \beta_0 a_1)}{\zeta} + O(|\zeta|^{-\frac{5}{3}}). \tag{28}$$

The area of $\Delta'(R)$ in the Q-metric is bounded above by the area of the domain on the w-sphere bounded by the image of $\gamma(R)$ under f (not containing the point at infinity). The difference of this quantity from the area of $\Delta(R)$ is given by

$$\Re\left\{ \frac{1}{2i} \int (\bar{\zeta} + \bar{\lambda}\bar{\zeta}^{-\frac{1}{3}} + \bar{\mu}\bar{\zeta}^{-1})(1 - \tfrac{1}{3}\lambda\zeta^{-\frac{4}{3}} - \mu\zeta^{-2}) d\zeta - \frac{1}{2i} \int \bar{\zeta} d\zeta \right\} + O(R^{-\frac{1}{3}})$$

where we have written λ for $(\frac{2}{3})^{\frac{1}{3}}\alpha^{\frac{2}{3}}a_1$, μ for $\frac{2}{3}\alpha(a_2 + \beta_0 a_1)$ and the integrals are taken over the three semicircles respectively in E_1, E_2, and E_3. The above expression is equal to

$$\Re\left\{\frac{1}{2i}\int_0^{3\pi}(Re^{-i\theta} + \bar{\lambda}R^{-\frac{1}{3}}e^{i\theta/3} + \bar{\mu}R^{-1}e^{i\theta})(1 - \tfrac{1}{3}\lambda R^{-\frac{4}{3}}e^{-4i\theta/3}\right.$$

$$\left. - \mu R^{-2}e^{-2i\theta})Re^{i\theta}id\theta - \frac{1}{2i}\int_0^{3\pi} Re^{-i\theta}Re^{i\theta}id\theta\right\} + O(R^{-\frac{1}{3}}).$$

An elementary calculation shows the whole expression to be $O(R^{-\frac{1}{3}})$. Thus

$$\iint\limits_{\Delta'(R)} dA \leq \iint\limits_{\Delta(R)} dA + O(R^{-\frac{1}{3}}) \tag{29}$$

where dA again denotes the element of area in the Q-metric.

To estimate the area of $\Delta'(R)$ from below we interpret it as the area of $\Delta(R)$ in a new metric

$$\rho(w)|Q(w)|^{\frac{1}{2}}|dw| = |f'(w)||Q(w)|^{\frac{1}{2}}|dw|$$

and apply the method of the extremal metric. We denote the intersections of $\mathfrak{E}_1$, $\mathfrak{E}_2$, $\mathfrak{E}_3$ with $\Delta(R)$ by $\mathfrak{E}_1(R)$, $\mathfrak{E}_2(R)$, $\mathfrak{E}_3(R)$ and the images of the latter under the given determination of $\int(Q(w))^{\frac{1}{2}}dw$ by $E_1(R)$, $E_2(R)$, $E_3(R)$. We transfer the above metric to the Riemann surface above the ζ-plane by setting

$$\rho(\zeta)|d\zeta| = \rho(w)|Q(w)|^{\frac{1}{2}}|dw|.$$

No confusion arises from using the notation ρ in each case.

Let now $\sigma(\theta)$ denote the horizontal segment lying in $E_1(R)$ on the line $\Im\zeta = R \sin\theta$. This is defined for all but perhaps one value of θ, $0 < \theta < \dfrac{\pi}{2}$. As in the proof of Theorem 1 we find

$$\int_{\sigma(\theta)} \rho d\xi \geq 2R\cos\theta + \Re\{\lambda R^{-\frac{1}{3}}(e^{-i\theta/3} - e^{-i(\pi-\theta)/3})\}$$

$$+ 2\Re\mu R^{-1}\cos\theta + O(R^{-\frac{4}{3}})$$

where λ and μ have the same meanings as above. Multiplying by $R\cos\theta$ and integrating with respect to θ over the range $\left(0, \dfrac{\pi}{2}\right)$ we obtain

$$\iint\limits_{E_1(R)} \rho d\xi d\eta \geq \frac{\pi}{2}R^2 + \frac{\pi}{2}\Re\mu$$

$$+ R^{\frac{2}{3}}\Re\left\{\lambda\int_0^{\frac{1}{2}\pi}(e^{-i\theta/3} - e^{-i(\pi-\theta)/3})\cos\theta d\theta\right\} + O(R^{-\frac{1}{3}}).$$

This can be expressed as

$$\iint\limits_{\mathfrak{E}_1(R)} \rho\, dA \geq \iint\limits_{\mathfrak{E}_1(R)} dA + \frac{\pi}{2}\,\Re\mu$$

$$+ R^{\frac{2}{3}}\Re\left\{\lambda \int_0^{\frac{1}{2}\pi} (e^{-i\theta/3} - e^{-i(\pi-\theta)/3}) \cos\theta\, d\theta\right\} + O(R^{-\frac{1}{3}}).$$

Similarly we have

$$\iint\limits_{\mathfrak{E}_2(R)} \rho\, dA \geq \iint\limits_{\mathfrak{E}_2(R)} dA + \frac{\pi}{2}\,\Re\mu$$

$$+ R^{\frac{2}{3}}\Re\left\{\lambda \int_\pi^{\frac{3}{2}\pi} (e^{-i\theta/3} - e^{-i(\pi-\theta)/3}) \cos\theta\, d\theta\right\} + O(R^{-\frac{1}{3}})$$

and

$$\iint\limits_{\mathfrak{E}_3(R)} \rho\, dA \geq \iint\limits_{\mathfrak{E}_3(R)} dA + \frac{\pi}{2}\,\Re\mu$$

$$+ R^{\frac{2}{3}}\Re\left\{\lambda \int_{2\pi}^{\frac{5}{2}\pi} (e^{-i\theta/3} - e^{-i(\pi-\theta)/3}) \cos\theta\, d\theta\right\} + O(R^{-\frac{1}{3}}).$$

Combining these results we find

$$(30) \quad \iint\limits_{\Delta(R)} \rho\, dA \geq \iint\limits_{\Delta(R)} dA + \tfrac{3}{2}\pi\,\Re\mu$$

$$+ R^{\frac{2}{3}}\Re\left\{\lambda\left[\int_0^{\frac{1}{2}\pi} + \int_\pi^{\frac{3}{2}\pi} + \int_{2\pi}^{\frac{5}{2}\pi}\right](e^{-i\theta/3} - e^{-i(\pi-\theta)/3}) \cos\theta\, d\theta\right\} + O(R^{-\frac{1}{3}}).$$

The third term on the right hand side can be rewritten as

$$R^{\frac{2}{3}}\Re\left\{\lambda \int_0^{\frac{1}{2}\pi} (e^{-i\theta/3} - e^{-i\pi/3}e^{i\theta/3} - e^{-i\pi/3}e^{-i\theta/3}\right.$$

$$\left. + e^{i\theta/3} + e^{-(2\pi/3)i}e^{-i\theta/3} - e^{i\pi/3}e^{i\theta/3}) \cos\theta\, d\theta\right\}$$

and the term in the inner bracket is seen to be identically zero, thus so is the entire term in question.

Using now the inequality

$$0 \leq \iint\limits_{\Delta(R)} (\rho - 1)^2 dA = \iint\limits_{\Delta(R)} \rho^2 dA + \iint\limits_{\Delta(R)} dA - 2\iint\limits_{\Delta(R)} \rho\, dA$$

we obtain

$$\iint_{\Delta'(R)} dA = \iint_{\Delta(R)} \rho^2 dA \geq 2\iint_{\Delta(R)} \rho dA - \iint_{\Delta(R)} dA$$

$$\geq \iint_{\Delta(R)} dA + 2\pi\Re\{\alpha(a_2 + \beta_0 a_1)\} + O(R^{-\frac{1}{2}}).$$

Comparing this with inequality (29) we obtain at once inequality (27).

To justify the equality statement we observe that unless f carries every trajectory of $Q(w)dw^2$ into another such it follows by the standard method that a positive constant can be inserted on the right hand side of inequality (30). The latter would imply at once that inequality (27) is strict. However in view of the expansion (28) the first alternative is possible only if f is the identity. Up to this point we have assumed the constant a_0 in the expansion of f about the point at infinity to be zero. Dropping now this assumption we see that equality can occur in (27) only if the mapping under f is a (Euclidean) translation.

9. Lemma 3. *Let $Q'_\phi(w)dw^2$ be the quadratic differential $e^{-3i\phi}wdw^2$. Then for*

$$-1 \leq \sigma \leq \tfrac{1}{3}$$

there exists a function $g(z, \sigma, \phi)$ in Σ mapping $|z| > 1$ onto an admissible domain Δ'_ϕ with respect to $Q'_\phi(w)dw^2$ bounded by segments emanating from the origin along the rays of argument ϕ, $\phi + \frac{2}{3}\pi$ and $\phi + \frac{4}{3}\pi$ (the latter two segments having equal length) so that the points $e^{i\phi}$, $e^{i(\phi+\theta)}$ and $e^{i(\phi-\theta)}$ on $|z| = 1$ correspond respectively to the other end points of these segments where θ, $0 < \theta \leq \pi$, satisfies

$$\sigma = 1 + 2\cos\theta.$$

This function is given explicitly by the expression

$$g(z, \sigma, \phi) = z(1 - 3\sigma e^{i\phi}z^{-1} - 3\sigma e^{2i\phi}z^{-2} + e^{3i\phi}z^{-3})^{\frac{2}{3}}$$

for a suitable choice of determination of the root and its expansion about the point at infinity begins

$$z - 2\sigma e^{i\phi} - \frac{\sigma(2+\sigma)e^{2i\phi}}{z} + \frac{\frac{4}{3}(1 - 3\sigma^2 - 2\sigma^3)e^{3i\phi}}{z^2} + \cdots.$$

Consider first the case $\phi = 0$. Then for a suitable choice of determination the mapping

$$\zeta = \int_0^w w^{\frac{1}{2}} dw$$

carries the upper half w-plane onto a domain bounded by the positive real axis and the negative imaginary axis, the image domain containing the upper half ζ-plane. This mapping is given explicitly by the formula

$$\zeta = \tfrac{2}{3} w^{\frac{3}{2}} \tag{31}$$

where the root has its positive determination for w large and positive.

Next the mapping

$$\zeta = \int_{-1}^{z} (z-1)(z-e^{i\theta})(z-e^{-i\theta})dz,$$

where $0 < \theta \leq \pi$, carries the domain $\Im z > 0$, $|z| > 1$ onto a domain bounded by the positive real axis, the negative imaginary axis and the segment

$$0 \geq \zeta \geq \tfrac{4}{3} \cos \tfrac{3}{2}\theta - 4(1 + 2 \cos \theta) \cos \tfrac{1}{2}\theta,$$

the image domain containing the upper half ζ-plane. This mapping is given explicitly by the formula

$$\zeta = \tfrac{2}{3} z^{\frac{3}{2}} - 2(1 + 2\cos\theta) z^{\frac{1}{2}} - 2(1 + 2\cos\theta) z^{-\frac{1}{2}} + \tfrac{2}{3} z^{-\frac{3}{2}} \tag{32}$$

where the roots have their positive determination for z large and positive. The image of the point $z = 1$ is $\zeta = \frac{4}{3} - 4(1 + 2 \cos \theta)$.

Provided that this latter point lies on the (closed) positive real axis the mappings (31) and (32) combine to give a conformal mapping from the domain $\Im z > 0$, $|z| > 1$ onto the upper half w-plane slit (except for $\theta = \pi$) along a segment of the ray of argument $\frac{2}{3}\pi$. By reflection this mapping extends to a conformal mapping $g(z, \sigma, 0)$ of $|z| > 1$ which is seen at once to be in Σ and to carry $|z| > 1$ onto the domain Δ_0'. Here σ stands for $1 + 2 \cos \theta$ and the above condition on the image of $z = 1$ is equivalent to requiring that

$$-1 \leq \sigma \leq \tfrac{1}{3}.$$

By equating the right hand sides of (31) and (32) we obtain at once the expression

$$g(z, \sigma, 0) = z(1 - 3\sigma z^{-1} - 3\sigma z^{-2} + z^{-3})^{\frac{2}{3}}$$

where the root has its positive determination for z large and positive. The expansion about the point at infinity is obtained directly.

Now to deal with the case of general real ϕ we form the functions

$$g(z, \sigma, \phi) = e^{i\phi} g(e^{-i\phi} z, \sigma, 0).$$

This completes the proof of Lemma 3.

10. Corollary 9. *Let $Q(w)dw^2$ be a quadratic differential on the w-sphere which is regular apart from a pole of order five at the point at infinity where Q has the expansion*

$$Q(w) = \alpha(w + \beta_0 + \text{terms in } w^{-1}).$$

Let $f^*(z) \in \Sigma$, map $|z| > 1$ onto a domain Δ admissible with respect to $Q(w)dw^2$ and have the expansion about the point at infinity

$$f^*(z) = z + b_0 + \frac{b_1}{z} + \frac{b_2}{z^2} + \dots .$$

Let $f(z) \in \Sigma$ and have the expansion (3) *about the point at infinity. Then*

$$\mathfrak{R}\{\alpha(c_2 + (b_0 + \beta_0)c_1)\} \leq \mathfrak{R}\{\alpha(b_2 + (b_0 + \beta_0)b_1)\}. \tag{33}$$

Equality can occur in (33) *only if*

$$f(z) \equiv f^*(z) + k,$$

k a constant.

Let $\psi(w)$ be the inverse function of $f^*(z)$ defined in Δ. Then Theorem 2 can be applied to the function

$$g(w) = f(\psi(w)).$$

The expansion of this function about the point at infinity is

$$g(w) = w + (c_0 - b_0) + \frac{(c_1 - b_1)}{w} + \frac{(b_0c_1 - b_0b_1 + c_2 - b_2)}{w^2} + \dots .$$

From inequality (27) then follows

$$\mathfrak{R}\{\alpha(c_2 - b_2 + b_0c_1 - b_0b_1 + \beta_0(c_1 - b_1)\} \leq 0$$

which is equivalent to inequality (33). The equality statement follows at once from Theorem 2.

COROLLARY 10. *Let $f(z) \in \Sigma$ and have the expansion* (3) *about the point at infinity. Then for ϕ real and $-1 \leq \sigma \leq \frac{1}{3}$*

$$\mathfrak{R}\{e^{-3i\phi}(c_2 - 2\sigma e^{i\phi}c_1)\} \leq \tfrac{2}{3} + 2\sigma^2 + \tfrac{4}{3}\sigma^3. \tag{34}$$

Equality can occur in (34) *only for the functions $g(z, \sigma, \phi) + k$, k a constant.*

We apply Corollary 9 with

$$Q(w)dw^2 = e^{-3i\phi}w\,dw^2$$

$$f^*(z) = g(z, \sigma, \phi)$$

so that

$$\begin{aligned} \alpha &= e^{-3i\phi} \\ \beta_0 &= 0 \\ b_0 &= -2\sigma e^{i\phi} \\ b_1 &= -\sigma(2 + \sigma)e^{2i\phi} \\ b_2 &= \tfrac{2}{3}(1 - 3\sigma^2 - 2\sigma^3)e^{3i\phi}. \end{aligned}$$

Inserting these values in inequality (33) we obtain inequality (34). The equality statement follows from that of Corollary 9.

Corollary 10 has various interesting more explicit consequences. However we confine ourselves here to giving the following coefficient result.

COROLLARY 11. *Let* $f(z) \in \Sigma$ *and have the expansion* (3) *about the point at infinity. Then*

(35) $$|c_2| \leq \tfrac{2}{3}.$$

Equality occurs in (35) *only for the functions* $z(1 + e^{3i\phi}z^{-3})^{\frac{2}{3}} + k$, ϕ *real,* k *a constant.*

Taking $\sigma = 0$ in inequality (34) we have for any real ϕ

$$\Re\{e^{-3i\phi}c_2\} \leq \tfrac{2}{3}.$$

We can choose ϕ so that

$$\Re\{e^{-3i\phi}c_2\} = |c_2|.$$

This gives inequality (35) and the equality statement follows by Corollary 10.

The result of this corollary is due to Golusin [4] and Schiffer [14].

11. THEOREM 3. *Let* $Q(w)dw^2$ *be a quadratic differential on the* w*-sphere which is regular apart from a pole of order six at the point at infinity where* Q *has the expansion*

$$Q(w) = \alpha(w^2 + \beta w + \beta_0 + \text{terms in } w^{-1}).$$

Let Δ *be an admissible domain with respect to* $Q(w)dw^2$ *and let* f *be a function univalent in* Δ *and regular apart from a simple pole at the point at infinity where it has the expansion*

$$f(w) = w + a_0 + \frac{a_1}{w} + \frac{a_2}{w^2} + \frac{a_3}{w^3} + \dots .$$

Then

(36) $$\Re\{\alpha[a_3 + \tfrac{1}{2}a_1^2 + \beta a_2 + \beta_0 a_1]\} \leq 0.$$

Equality can occur in (36) *only if the mapping under* f *is composed of a Euclidean translation and a translation along the trajectories of* $Q(w)dw^2$.

This result, like Theorem 1 and Theorem 2, admits numerous generalizations. The proof follows very closely that of Theorem 2 so that we may omit some details in presenting it.

As before we take a sufficiently small simply-connected neighborhood U of the point at infinity slit along a trajectory Λ of $Q(w)dw^2$. The image of the set $U - \overline{\Lambda}$ by a branch of $\zeta = \int(Q(w))^{\frac{1}{2}}dw$ will be a portion of Riemann surface over the ζ-plane. We take the trace on this surface of a circle centre the origin of sufficiently large radius R and its inverse image on the w-sphere. Joining the end points of the latter if necessary by an arc of Λ we obtain a simple closed curve $\gamma(R)$. Removing the closure of the neighborhood of the point at infinity bounded by $\gamma(R)$ from Δ we obtain a domain which we denote by $\Delta(R)$. We assume R so large that the boundary of Δ lies in the interior of the closure of $\Delta(R)$. The image of $\Delta(R)$ under f is denoted by $\Delta'(R)$.

The determination of $\int (Q(w))^{\frac{1}{2}} dw$ may be chosen so that we have for the mapping of $U - \bar{\Lambda}$

$$\zeta = \int \alpha^{\frac{1}{2}} w \left(1 + \frac{\beta}{w} + \frac{\beta_0}{w^2} + \frac{\beta_1}{w^3} + \frac{\beta_2}{w^4} + \dots\right)^{\frac{1}{2}} dw$$

$$= \tfrac{1}{2}\alpha^{\frac{1}{2}} w^2 + \tfrac{1}{2}\alpha^{\frac{1}{2}}\beta w + (\tfrac{1}{2}\alpha^{\frac{1}{2}}\beta_0 - \tfrac{1}{8}\alpha^{\frac{1}{2}}\beta^2)\log w - \alpha^{\frac{1}{2}}(\tfrac{1}{2}\beta_1 - \tfrac{1}{4}\beta\beta_0 + \tfrac{1}{16}\beta^3)w^{-1}$$
$$- \tfrac{1}{2}\alpha^{\frac{1}{2}}(\tfrac{1}{2}\beta_2 - \tfrac{1}{4}\beta\beta_1 - \tfrac{1}{8}\beta_0^2 + \tfrac{3}{16}\beta^2\beta_0 - \tfrac{5}{128}\beta^4)w^{-2} + \text{higher powers of } w^{-1}.$$

Again the mapping f in the w-sphere induces a mapping in terms of the variable ζ and we denote by ω the value corresponding to ζ. We may assume for the moment that the coefficient a_0 in the expansion of $f(w)$ about the point at infinity is zero. Then for ζ large enough

$$\omega = \tfrac{1}{2}\alpha^{\frac{1}{2}}\left(w + \frac{a_1}{w} + \frac{a_2}{w^2} + \frac{a_3}{w^3} + \dots\right)^2$$
$$+ \tfrac{1}{2}\alpha^{\frac{1}{2}}\beta\left(w + \frac{a_1}{w} + \frac{a_2}{w^2} + \frac{a_3}{w^3} + \dots\right)$$
$$+ (\tfrac{1}{2}\alpha^{\frac{1}{2}}\beta_0 - \tfrac{1}{8}\alpha^{\frac{1}{2}}\beta^2)\log\left(w + \frac{a_1}{w} + \frac{a_2}{w^2} + \frac{a_3}{w^3} + \dots\right)$$
$$- \alpha^{\frac{1}{2}}(\tfrac{1}{2}\beta_1 - \tfrac{1}{4}\beta\beta_0 + \tfrac{1}{16}\beta^3)\left(w + \frac{a_1}{w} + \dots\right)^{-1}$$
$$- \tfrac{1}{2}\alpha^{\frac{1}{2}}(\tfrac{1}{2}\beta_2 - \tfrac{1}{4}\beta\beta_1 - \tfrac{1}{8}\beta_0^2 + \tfrac{3}{16}\beta^2\beta_0 - \tfrac{5}{128}\beta^4)\left(w + \frac{a_1}{w} + \dots\right)^{-2} + \dots$$
$$= \zeta + \alpha^{\frac{1}{2}}a_1 + \alpha^{\frac{1}{2}}(a_2 + \tfrac{1}{2}\beta a_1)w^{-1} + \alpha^{\frac{1}{2}}(a_3 + \tfrac{1}{2}a_1^2$$
$$+ \tfrac{1}{2}\beta a_2 + \tfrac{1}{2}\beta_0 a_1 - \tfrac{1}{8}\beta^2 a_1)w^{-2} + \dots .$$

On the other hand

$$w^{-2} = \tfrac{1}{2}\alpha^{\frac{1}{2}}\zeta^{-1}(1 + \beta w^{-1} + (\beta_0 - \tfrac{1}{4}\beta^2)w^{-2}\log w + \dots)$$

and

$$w^{-1} = 2^{-\frac{1}{2}}\alpha^{\frac{1}{4}}\zeta^{-\frac{1}{2}}(1 + \beta w^{-1} + \dots)^{\frac{1}{2}}$$
$$= 2^{-\frac{1}{2}}\alpha^{\frac{1}{4}}\zeta^{-\frac{1}{2}} + \tfrac{1}{4}\alpha^{\frac{1}{2}}\beta\zeta^{-1} + \dots$$

with appropriate choices of the roots. Thus finally

$$\omega = \zeta + \alpha^{\frac{1}{2}}a_1 + \frac{2^{-\frac{1}{2}}\alpha^{\frac{3}{4}}(a_2 + \frac{1}{2}\beta a_1)}{\zeta^{\frac{1}{2}}} \tag{37}$$
$$+ \frac{\frac{1}{2}\alpha\,(a_3 + \frac{1}{2}a_1^2 + \beta a_2 + \frac{1}{2}\beta_0 a_1 + \frac{1}{8}\beta^2 a_1)}{\zeta} + O(|\zeta|^{-\frac{3}{2}}).$$

The area of $\Delta'(R)$ in the Q-metric is bounded above by the area of the domain on the w-sphere bounded by the image of $\gamma(R)$ under f (not containing the point at infinity). The change in area may be regarded as arising from the effect of the mapping on the circumference of a circle described twice adjusted by the effect of possibly a horizontal segment and a circular arc. The former contribution is seen at once to be

$$\mathfrak{R}\left\{\frac{1}{2i}\int(\bar{\zeta}+\bar{\kappa}+\bar{\lambda}\bar{\zeta}^{-\frac{1}{2}}+\bar{\mu}\bar{\zeta}^{-1})(1-\tfrac{1}{2}\lambda\zeta^{-\frac{3}{2}}-\mu\zeta^{-2})d\zeta-\frac{1}{2i}\int\bar{\zeta}d\zeta\right\}+O(R^{-\frac{1}{2}})$$

where we have written κ for $\alpha^{\frac{1}{2}}a_1$, λ for $2^{-\frac{1}{2}}\alpha^{\frac{3}{4}}(a_2+\frac{1}{2}\beta a_1)$, μ for $\frac{1}{2}\alpha(a_3+\frac{1}{2}a_1^2+\beta a_2+\frac{1}{2}\beta_0 a_1+\frac{1}{8}\beta^2 a_1)$ and the integrals are taken over the circumference described twice. The above expression is equal to

$$\mathfrak{R}\left\{\frac{1}{2i}\int_0^{4\pi}(Re^{-i\theta}+\bar{\kappa}+\bar{\lambda}R^{-\frac{1}{2}}e^{\frac{1}{2}i\theta}+\bar{\mu}Re^{i\theta})(1-\tfrac{1}{2}\lambda R^{-\frac{3}{2}}e^{-\frac{3}{2}i\theta}\right.$$
$$\left.-\mu R^{-2}e^{-2i\theta})Re^{i\theta}id\theta-\frac{1}{2i}\int_0^{4\pi}Re^{-i\theta}Re^{i\theta}id\theta\right\}+O(R^{-\frac{1}{2}}).$$

An elementary calculation shows the whole expression to be $O(R^{-\frac{1}{2}})$. A circuit in the z-sphere about the point at infinity returns us in the Riemann surface over the ζ-plane to a point whose affix has been translated by $\pi i\alpha^{\frac{1}{2}}(\beta_0-\frac{1}{4}\beta^2)$. Since up to terms which are $O(R^{-\frac{1}{2}})$ the transformation from ζ to ω is a translation through $\alpha^{\frac{1}{2}}a_1$ the contribution from the horizontal segment and the circular arc is seen to be

$$\mathfrak{I}\{\pi i\alpha^{\frac{1}{2}}(\beta_0-\tfrac{1}{4}\beta^2)\bar{\alpha}^{\frac{1}{2}}\bar{a}_1\}+O(R^{-\frac{1}{2}})=\pi\mathfrak{R}\{|\alpha|(\beta_0-\tfrac{1}{4}\beta^2)\bar{a}_1\}+O(R^{-\frac{1}{2}}).$$

Thus the total change in area is

$$\pi\mathfrak{R}\{|\alpha|(\beta_0-\tfrac{1}{4}\beta^2)\bar{a}_1\}+O(R^{-\frac{1}{2}})$$

and we have

$$\iint\limits_{\Delta'(R)}dA\leq\iint\limits_{\Delta(R)}dA+\pi\mathfrak{R}\{|\alpha|(\beta_0-\tfrac{1}{4}\beta^2)\bar{a}_1\}+O(R^{-\frac{1}{2}})\tag{38}$$

where dA denotes the element of area in the Q-metric.

To estimate the area of $\Delta'(R)$ from below we interpret it as the area of $\Delta(R)$ in the new metric

$$\rho(w)|Q(w)|^{\frac{1}{2}}|dw|=|f'(w)||Q(w)|^{\frac{1}{2}}|dw|$$

and apply the method of the extremal metric. In the present case it is readily seen that the domains in the trajectory structure of $Q(w)dw^2$ are four end domains $\mathfrak{E}_1$, $\mathfrak{E}_2$, $\mathfrak{E}_3$, $\mathfrak{E}_4$ which we may suppose occur in cyclic order about the point at infinity and a possible strip domain $\mathfrak{S}$. We denote the respective intersections of these domains with $\Delta(R)$ by $\mathfrak{E}_1(R)$, $\mathfrak{E}_2(R)$, $\mathfrak{E}_3(R)$, $\mathfrak{E}_4(R)$, and $\mathfrak{S}(R)$. With an appropriate numbering of the end domains and λ and μ having the same meanings as above we readily verify that

$$\iint\limits_{\mathfrak{E}_1(R)} \rho\, dA \geq \iint\limits_{\mathfrak{E}_1(R)} dA + \frac{\pi}{2}\Re\mu$$
$$+ R^{\frac{1}{2}}\Re\left\{\lambda\int_0^{\frac{1}{2}\pi}(e^{-i(\theta/2)} - e^{-\frac{1}{2}i(\pi-\theta)})\cos\theta\, d\theta\right\} + O(R^{-\frac{1}{2}})$$

$$\iint\limits_{\mathfrak{E}_2(R)} \rho\, dA \geq \iint\limits_{\mathfrak{E}_2(R)} dA + \frac{\pi}{2}\Re\mu$$
$$+ R^{\frac{1}{2}}\Re\left\{\lambda\int_{\pi}^{\frac{3}{2}\pi}(e^{-i(\theta/2)} - e^{-\frac{1}{2}i(\pi-\theta)})\cos\theta\, d\theta\right\} + O(R^{-\frac{1}{2}})$$

$$\iint\limits_{\mathfrak{E}_3(R)} \rho\, dA \geq \iint\limits_{\mathfrak{E}_3(R)} dA + \frac{\pi}{2}\Re\mu$$
$$+ R^{\frac{1}{2}}\Re\left\{\lambda\int_{2\pi}^{\frac{5}{2}\pi}(e^{-i(\theta/2)} - e^{-\frac{1}{2}i(\pi-\theta)})\cos\theta\, d\theta\right\} + O(R^{-\frac{1}{2}})$$

$$\iint\limits_{\mathfrak{E}_4(R)} \rho\, dA \geq \iint\limits_{\mathfrak{E}_4(R)} dA + \frac{\pi}{2}\Re\mu$$
$$+ R^{\frac{1}{2}}\Re\left\{\lambda\int_{3\pi}^{\frac{7}{2}\pi}(e^{-i(\theta/2)} - e^{-\frac{1}{2}i(\pi-\theta)})\cos\theta\, d\theta\right\} + O(R^{-\frac{1}{2}}).$$

If a strip domain $\mathfrak{S}$ is present, it is verified that

$$\iint\limits_{\mathfrak{S}(R)} \rho\, dA \geq \iint\limits_{\mathfrak{S}(R)} dA + \Re\{\pi\alpha^{\frac{1}{2}}(\beta_0 - \tfrac{1}{4}\beta^2)\}\Re\{\alpha^{\frac{1}{2}}a_1\} + O(R^{-\frac{1}{2}}).$$

Combining these results we find

$$\text{(39)}\quad \iint\limits_{\Delta(R)} \rho\, dA \geq \iint\limits_{\Delta(R)} dA + 2\pi\Re\mu + \Re\{\pi\alpha^{\frac{1}{2}}(\beta_0 - \tfrac{1}{4}\beta^2)\}\Re\{\alpha^{\frac{1}{2}}a_1\}$$
$$+ R^{\frac{1}{2}}\Re\left\{\lambda\left[\int_0^{\frac{1}{2}\pi} + \int_{\pi}^{\frac{3}{2}\pi} + \int_{2\pi}^{\frac{5}{2}\pi} + \int_{3\pi}^{\frac{7}{2}\pi}\right](e^{-\frac{1}{2}i\theta} - e^{-\frac{1}{2}i(\pi-\theta)})\cos\theta\, d\theta\right\} + O(R^{-\frac{1}{2}}).$$

The fourth term on the right hand side can be rewritten as

$$R^{\frac{1}{2}}\Re\left\{\lambda\int_0^{\pi/2}(e^{-i(\theta/2)} - e^{-\frac{1}{2}i\pi}e^{i(\theta/2)} - e^{-\frac{1}{2}i\pi}e^{-i(\theta/2)}\right.$$
$$\left. + e^{i(\theta/2)} - e^{-i(\theta/2)} - e^{\frac{1}{2}i\pi}e^{i(\theta/2)} - e^{-3i\pi/2}e^{-i(\theta/2)} - e^{i(\theta/2)})\cos\theta\, d\theta\right\}$$

and the term in the inner bracket is seen to be identically zero, thus so is the entire term in question.

Using now the inequality

$$0 \leq \iint\limits_{\Delta(R)} (\rho - 1)^2 dA = \iint\limits_{\Delta(R)} \rho^2 dA + \iint\limits_{\Delta(R)} dA - 2\iint\limits_{\Delta(R)} \rho\, dA$$

we obtain

$$\iint\limits_{\Delta'(R)} dA = \iint\limits_{\Delta(R)} \rho^2 dA \geq 2\iint\limits_{\Delta(R)} \rho dA - \iint\limits_{\Delta(R)} dA$$

$$\geq \iint\limits_{\Delta(R)} dA + 2\pi\Re\{\alpha(a_3 + \tfrac{1}{2}a_1^2 + \beta a_2 + \tfrac{1}{2}\beta_0 a_1 + \tfrac{1}{8}\beta^2 a_1)\}$$

$$+ 2\Re\{\pi\alpha^{\frac{1}{2}}(\beta_0 - \tfrac{1}{4}\beta^2)\}\Re\{\alpha^{\frac{1}{2}}a_1\} + O(R^{-\frac{1}{2}}).$$

Comparing this with inequality (38) we obtain at once inequality (36).

To justify the equality statement we observe that unless f carries every trajectory of $Q(w)dw^2$ into another such it follows by the standard method that a positive constant can be inserted on the right hand side of inequality (39). The latter would imply at once that inequality (36) is strict. On the other hand, since in the present situation $Q(w)dw^2$ must have zeros at finite points, the first alternative would require f to carry each trajectory of $Q(w)dw^2$ into itself. Thus the mapping under f would have to be a translation (in the Q-metric) along trajectories of $Q(w)dw^2$. Up to this point we have assumed the constant a_0 in the expansion of f about the point at infinity to be zero. Dropping now this assumption we see that equality can occur in (36) only if the mapping under f is composed of a Euclidean translation and a translation along the trajectories of $Q(w)dw^2$.

12. LEMMA 4. *Let $Q''_\psi(w)dw^2$ be the quadratic differential $e^{-4i\psi}(\sigma^2 e^{2i\psi} - w^2)dw^2$ (ψ real, $0 \leq \sigma \leq 2$). Let $h(z, \sigma, \psi)$ be the function in Σ mapping $|z| > 1$ conformally onto the domain Δ''_ψ bounded by the rectilinear segment joining the zeros $\sigma e^{i\psi}$, $-\sigma e^{i\psi}$ of $Q''_\psi(w)dw^2$ together with four arcs of equal length on the other trajectories of $Q''_\psi(w)dw^2$ having limiting end points at these zeros (when $\sigma = 0$ the zeros coincide and the first segment is missing; when $\sigma = 2$ the latter four arcs are missing). For $\sigma = 0$ we have*

$$h(z, 0, \psi) = z(1 - e^{4i\psi}z^{-4})^{\frac{1}{4}}$$

the root being chosen so that the expansion of $h(z, 0, \psi)$ about the point at infinity begins

$$z - \tfrac{1}{4}e^{4i\psi}z^{-3} + \dots .$$

The functions which can be obtained from $h(z, 0, \psi)$ by translation along the trajectories of $Q''_\psi(w)dw^2$ are a one parameter family $H(z, 0, \psi, \mu)$ where

$$H(z, 0, \psi, \mu) = z(1 + 2i\mu e^{2i\psi}z^{-2} - e^{4i\psi}z^{-4})^{\frac{1}{2}},$$

the root is chosen so that the expansion of $H(z, 0, \psi, \mu)$ about the point at infinity begins

$$z + i\mu e^{2i\psi}z^{-1} + \tfrac{1}{2}(\mu^2 - 1)e^{4i\psi}z^{-3} + \dots,$$

and μ $(-1 \leq \mu \leq 1)$ denotes the (sensed) amount of translation in the

Q''_ψ-metric. For $0 < \sigma \leq 2$ the expansion of $h(z, \sigma, \psi)$ about the point at infinity begins

$$z + \tfrac{1}{4}\sigma^2[1 - \log(\sigma^2/4)]e^{2i\psi}z^{-1}$$
$$+ [\tfrac{1}{32}\sigma^4 - \tfrac{1}{16}\sigma^4 \log(\sigma^2/4) - \tfrac{1}{32}\sigma^4[\log(\sigma^2/4)]^2 - \tfrac{1}{2}]e^{4i\psi}z^{-3} + \dots .$$

When $0 < \sigma < 2$ there can be obtained from $h(z, \sigma, \psi)$ by translation along the trajectories of $Q''_\psi(w)dw^2$ a one parameter family of functions $H(z, \sigma, \psi, u)$ where μ denotes the (sensed) amount of translation and satisfies the inequalities

$$-(1 - \tfrac{1}{16}\sigma^4)^{\frac{1}{2}} + \tfrac{1}{4}\sigma^2 \cos^{-1}\frac{\sigma^2}{4} \leq \mu \leq (1 - \tfrac{1}{16}\sigma^4)^{\frac{1}{2}} - \tfrac{1}{4}\sigma^2 \cos^{-1}\frac{\sigma^2}{4}.$$

In particular

$$H(z, \sigma, \psi, 0) \equiv h(z, \sigma, \psi).$$

The expansion of $H(z, \sigma, \psi, \mu)$ about the point at infinity begins

$$z + \tfrac{1}{2}[2i\mu + \tfrac{1}{2}\sigma^2(1 - \log(\sigma^2/4))]e^{2i\psi}z^{-1}$$
$$+ [\tfrac{1}{2}i\mu\,\sigma^2 + \tfrac{1}{2}\mu^2 - \tfrac{1}{4}i\mu\sigma^2(1 - \log(\sigma^2/4)) + \tfrac{1}{32}\sigma^4$$
$$- \tfrac{1}{16}\sigma^4 \log(\sigma^2/4) - \tfrac{1}{32}\sigma^4(\log(\sigma^2/4))^2 - \tfrac{1}{2}]e^{4i\psi}\,z^{-3} + \dots .$$

For $\sigma = 2$ there are no functions obtained from $h(z, 2, \psi)$ by translation along the trajectories of $Q''_\psi(w)dw^2$.

For $\sigma = 0$ the result of the lemma is elementary. For $0 < \sigma \leq 2$, recalling the notation of Lemma 1, if we set $W = w^2$, $Z = z^2$, $\tau = \sigma^2$, $\phi = 2\psi$, $\lambda = 2\mu$, we readily find

$$Q_\phi(W)dW^2 = 4Q''_\psi(w)dw^2,$$
$$[f(Z, \tau, \phi)]^{\frac{1}{2}} = h(z, \sigma, \psi),$$
$$[F(Z, \tau, \phi, \lambda)]^{\frac{1}{2}} = H(z, \sigma, \psi, \mu)$$

with appropriate choices of the roots. From this the remaining results of lemma follow directly.

13. Corollary 12. *Let $Q(w)dw^2$ be a quadratic differential on the w-sphere which is regular apart from a pole of order six at the point at infinity where Q has the expansion*

$$Q(w) = \alpha(w^2 + \beta w + \beta_0 + \text{terms in } w^{-1}).$$

Let $f^*(z) \in \Sigma$, map $|z| > 1$ onto a domain Δ admissible with respect to $Q(w)dw^2$ and have the expansion about the point at infinity

$$f^*(z) = z + b_0 + \frac{b_1}{z} + \frac{b_2}{z^2} + \frac{b_3}{z^3} + \dots .$$

Let $f(z) \in \Sigma$ and have the expansion (3) about the point at infinity. Then

$$(40) \quad \Re\{\alpha(c_3 + \tfrac{1}{2}c_1^2 + 2b_0c_2 + c_1b_0^2 + \beta c_2 + \beta c_1b_0 + \beta_0c_1)\}$$
$$\leq \Re\{\alpha(b_3 + \tfrac{1}{2}b_1^2 + 2b_0b_2 + b_1b_0^2 + \beta b_2 + \beta b_1b_0 + \beta_0b_1)\}.$$

Equality can occur in (40) *only if* $f(z)$ *is obtained from* $f^*(z)$ *by a translation along the trajectories of* $Q(w)dw^2$ *together with a Euclidean translation.*

Let $\psi(w)$ be the inverse function of $f^*(z)$ defined in Δ. Then Theorem 3 can be applied to the function

$$g(w) = f(\psi(w)).$$

The expansion of this function about the point at infinity is

$$g(w) = w + (c_0 - b_0) + \frac{(c_1 - b_1)}{w} + \frac{(c_2 - b_2 + c_1b_0 - b_1b_0)}{w^2}$$
$$+ \frac{(c_3 - b_3 + 2b_0c_2 - 2b_0b_2 + c_1b_1 - b_1^2 + c_1b_0^2 - b_1b_0^2)}{w^3} + \cdots.$$

From inequality (36) then follows

$$\Re\{\alpha[c_3 - b_3 + 2b_0c_2 - 2b_0b_2 + c_1b_1 - b_1^2 + c_1b_0^2 - b_1b_0^2 + \tfrac{1}{2}(c_1 - b_1)^2 + \beta(c_2 - b_2 + c_1b_0 - b_1b_0) + \beta_0(c_1 - b_1)]\} \leq 0,$$

which is equivalent to inequality (40). The equality statement follows at once from Theorem 3.

COROLLARY 13. *Let* $f(z) \in \Sigma$ *and have the expansion* (3) *about the point at infinity. Then for* ψ *real and* $0 \leq \sigma \leq 2$

$$(41) \quad \Re\{e^{-4i\psi}(c_3 + \tfrac{1}{2}c_1^2 - \sigma^2e^{2i\psi}c_1)\} \geq -\tfrac{1}{2} - \tfrac{3}{16}\sigma^4 + \tfrac{1}{8}\sigma^4 \log(\sigma^2/4); \quad 0 < \sigma \leq 2$$
$$\geq -\tfrac{1}{2} \quad , \sigma = 0.$$

For $0 \leq \sigma < 2$ *equality occurs in* (41) *only for the functions* $H(z, \sigma, \psi, \mu) + k$, k *a constant. For* $\sigma = 2$ *equality occurs in* (41) *only for the functions* $h(z, \sigma, \psi) + k$, k *a constant.*

We apply Corollary 12 with

$$Q(w)dw^2 = e^{-4i\psi}(\sigma^2e^{2i\psi} - w^2)dw^2$$
$$f^*(z) = h(z, \sigma, \psi)$$

so that

$$\alpha = -e^{-4i\psi}$$
$$\beta = 0$$
$$\beta_0 = -\sigma^2e^{2i\psi}$$
$$b_0 = 0$$
$$b_1 = 0, \quad \sigma = 0$$
$$= \tfrac{1}{4}\sigma^2[1 - \log(\sigma^2/4)], \quad 0 < \sigma \leq 2$$
$$b_2 = 0$$
$$b_3 = -\tfrac{1}{2}e^{4i\psi}, \quad \sigma = 0$$
$$= [\tfrac{1}{32}\sigma^4 - \tfrac{1}{16}\sigma^4 \log(\sigma^2/4) - \tfrac{1}{32}\sigma^4[\log(\sigma^2/4)]^2 - \tfrac{1}{2}]e^{4i\psi}, \quad 0 < \sigma \leq 2.$$

Inserting these values in inequality (40) we obtain inequality (41). The equality statements follow from that of Corollary 12 and the enumeration in Lemma 4 of the functions which can be obtained from $h(z, \sigma, \psi)$ by translation along the trajectories of $Q''_{\psi}(w)dw^2$.

Corollary 13 has various interesting more explicit consequences. However we confine ourselves here to giving the following coefficient result.

COROLLARY 14. *Let $f(z) \in \Sigma$ and have the expansion* (3) *about the point at infinity. Then*

$$|c_3| \leq \tfrac{1}{2} + e^{-6}. \tag{42}$$

Equality occurs in (42) *only for the functions $h(z, 2e^{-\frac{3}{2}}, \psi) + k$, ψ real, k a constant.*

We can choose ψ real so that

$$-\Re\{e^{-4i\psi}c_3\} = |c_3|$$

while

$$\Re\{e^{-2i\psi}c_1\} \geq 0.$$

Then from inequality (41) follows for $0 \leq \sigma \leq 2$

$$\begin{aligned} (43)\quad |c_3| &\leq \tfrac{1}{2} + \tfrac{3}{16}\sigma^4 + \tfrac{1}{8}\sigma^4 \log(\sigma^2/4) + \tfrac{1}{2}\Re\{e^{-4i\psi}c_1^2\} - \sigma^2\Re\{e^{-2i\psi}c_1\}, \\ & \qquad\qquad\qquad\qquad\qquad\qquad\qquad\qquad\qquad\qquad 0 < \sigma \leq 2 \\ &\leq \tfrac{1}{2} + \tfrac{1}{2}\Re\{e^{-4i\psi}c_1^2\}, \qquad\qquad\qquad\qquad\qquad \sigma = 0. \end{aligned}$$

Now

$$\Re\{e^{-4i\psi}c_1^2\} \leq [\Re\{e^{-2i\psi}c_1\}]^2 \tag{44}$$

the inequality being strict unless

$$\Im\{e^{-2i\psi}c_1\} = 0.$$

If

$$\Re\{e^{-2i\psi}c_1\} = 0$$

taking $\sigma = 0$ in inequality (43) we have

$$|c_3| \leq \tfrac{1}{2}.$$

If

$$\Re\{e^{-2i\psi}c_1\} > 0$$

we can find σ, $0 < \sigma \leq 2$, so that

$$\Re\{e^{-2i\psi}c_1\} = \tfrac{1}{4}\sigma^2[1 - \log(\sigma^2/4)].$$

Using this value of σ we obtain from inequality (43)

$$|c_3| \leq \tfrac{1}{2} - \tfrac{1}{32}\sigma^4 + \tfrac{1}{16}\sigma^4 \log(\sigma^2/4) + \tfrac{1}{32}\sigma^4[\log(\sigma^2/4)]^2.$$

Now we define

$$\begin{aligned}\Theta(\sigma) &= \tfrac{1}{2} - \tfrac{1}{32}\sigma^4 + \tfrac{1}{16}\sigma^4 \log(\sigma^2/4) + \tfrac{1}{32}\sigma^4[\log(\sigma^2/4)]^2, \quad 0 < \sigma \leq 2 \\ &= \tfrac{1}{2} \qquad\qquad\qquad\qquad\qquad\qquad\qquad\qquad\qquad\quad , \quad \sigma = 0.\end{aligned}$$

The function $\Theta(\sigma)$ is continuous on $0 \leq \sigma \leq 2$. Further we have

$$|c_3| \leq \max_{0 \leq \sigma \leq 2} \Theta(\sigma).$$

Now for $0 < \sigma \leq 2$

$$\frac{d\Theta(\sigma)}{d\sigma} = \tfrac{1}{8}\sigma^3 \log(\sigma^2/4)[3 + \log(\sigma^2/4)].$$

Thus the function $\Theta(\sigma)$ increases from $\sigma = 0$ up to the place where

$$3 + \log(\sigma^2/4) = 0$$

that is

$$\sigma = 2e^{-\frac{3}{2}}$$

and then decreases. Hence the maximum of $\Theta(\sigma)$ occurs for this value of σ and

$$|c_3| \leq \tfrac{1}{2} + e^{-6}.$$

Since we have used Corollary 13 and inequality (44), equality can occur here only for the functions $h(z, 2e^{-\frac{3}{2}}, \psi) + k$, ψ real, k a constant.

14. The method illustrated here by these few examples has many other applications. Indeed there is a corresponding extension of the General Coefficient Theorem itself.

References

[1] L. Bieberbach, Über die Koeffizienten derjenigen Potenzreihen, welche eine schlichte Abbildung des Einheitskreises vermitteln, *Sitzungsberichte der Königlich Preussischen Akademie der Wissenschaften* (1916), zweiter Halbband, pp. 940–955.

[2] P. R. Garabedian and M. Schiffer, A coefficient inequality for schlicht functions, *Annals of Mathematics 61* (1955), pp. 116–136.

[3] ———, A proof of the Bieberbach conjecture for the fourth coefficient, *Journal of Rational Mechanics and Analysis 4* (1955), pp. 427–465.

[4] G. M. Golusin, Some evaluations of the coefficients of univalent functions, *Matematicheskii Sbornik 3* (new series) (1938), pp. 321–330 (Russian, German summary).

[5] ———, On distortion theorems and the coefficients of univalent functions, *ibidem 19* (new series) (1946), pp. 183–202 (Russian, English summary).

[6] ———, Some problems in the theory of univalent functions, *Trudy Matematicheskogo instituta imeni V. A. Steklova, No. 27* (1949) (Russian).

[7] James A. Jenkins, Remarks on "Some problems in conformal mapping", *Proceedings of the American Mathematical Society 3* (1952), pp. 147–151.

[8] ———, On the local structure of the trajectories of a quadratic differential, *ibidem 5* (1954), pp. 357–362.

[9] ———, A general coefficient theorem, *Transactions of the American Mathematical Society 77* (1954), pp. 262–280.

[10] ———, Univalent functions and conformal mapping, to appear in *the Ergebnisse series.*

[11] JAMES A. JENKINS and D. C. SPENCER, Hyperelliptic trajectories, *Annals of Mathematics 53* (1951), pp. 4–35.

[12] K. LÖWNER, Untersuchungen über schlichte konforme Abbildungen des Einheitskreises, I, *Mathematische Annalen 89* (1923), pp. 103–121.

[13] A. C. SCHAEFFER and D. C. SPENCER, Coefficient regions for schlicht functions, *American Mathematical Society Colloquium Publications 35* (1950).

[14] M. SCHIFFER, Sur un problème d'extrémum de la représentation conforme, *Bullétin de la Société Mathématique de France 65* (1937), pp. 48–55.

[15] O. TEICHMÜLLER, Ungleichungen zwischen den Koeffizienten schlichter Funktionen, *Sitzungsberichte der Preussischen Akademie der Wissenschaften*, Physikalisch-Mathematische Klasse, (1938), pp. 363–375.

Appendix

Contents of "Seminars on Analytic Functions"

Volume 1

APPENDIX

On the Phragmén–Lindelöf theorem and some applications	H. RADEMACHER
A division problem for analytic functions	P. C. ROSENBLOOM and S. E. WARSCHAWSKI
An extremal problem	F. SPRINGER

Volume 2

Seminar III. Riemann Surfaces

Abel's theorem for open Riemann surfaces	L. V. AHLFORS
Determination of an automorphic function for a given analytic equivalence relation	L. FOURÈS
A theorem concerning the existence of deformable conformal maps	M. HEINS
A direct construction of Abelian differentials on Riemann surfaces	A. PFLUGER
The first variations of the Douglas functional and the periods of the Abelian integrals of the first	H. E. RAUCH
Stability problems on boundary components	L. SARIO
Approximation by bounded analytic functions	J. L. WALSH

Seminar IV. Theory of Automorphic Functions

Induced representation	R. BOTT
Characteristic numbers of homogeneous domains	F. HIRZEBRUCH
Automorphic forms in half-spaces	M. KOECHER
The Fourier coefficients of automorphic forms belonging to a class of real zonal horocyclic groups	J. LEHNER
Asymptotic formulas for the Fourier coefficients of multiplicative automorphic functions	H. PETERSSON
Automorphic functions and integral operators	A. SELBERG

Seminar V. Analytic Functions as Related to Banach Algebras

Joint spectra in topological algebras	R. ARENS
Algebraic properties of classes of analytic functions	R. C. BUCK
Multiplicative functionals on Banach algebras	L. CARLESON
Linear inequalities and closure properties in normed linear spaces	K. FAN
Function algebras	A. M. GLEASON
Fatou's theorem for generalized analytic functions	K. HOFFMAN
On rings of bounded analytic functions	S. KAKUTANI
Derivations of Banach algebras	I. KAPLANSKY
The Laplace transform on groups and generalized analytic functions	G. W. MACKEY

APPENDIX

Rings of meromorphic functions	H. ROYDEN
Analytic functions on locally convex algebras	L. A. WAELBROECK
Rings of analytic functions	J. WERMER
Spectral decomposition of operators in a Banach space and the analytic character of their resolvent	F. WOLF